D1104512

Carbon, Nitrogen, and Sulfur Pollutants and Their Determination in Air and Water

JEROME GREYSON

Metropolitan State College, Denver
J & JG Associates, Conifer, Colorado

Marcel Dekker, Inc.   •   New York and Basel

Library of Congress Cataloging-in-Publication Data

Greyson, Jerome C.
   Carbon, nitrogen, and sulfur pollutants and their determination in
air and water / Jerome C. Greyson.
      p. cm.
   Includes bibliographical references and index.
   ISBN 0-8247-8235-6
   1. Air--Pollution--Measurement. 2. Water--Pollution--Measurement.
3. Carbon--Environmental aspects. 4. Nitrogen--Environmental
aspects. 5. Sulphur--Environmental aspects. I. Title.
TD890.G74   1990
628.5'3--dc20   90-41742
                                                                      CIP

This book is printed on acid-free paper.

MARCEL DEKKER, INC.
270 Madison Avenue, New York, New York 10016

Current printing (last digit):
10 9 8 7 6 5 4 3 2 1

PRINTED IN THE UNITED STATES OF AMERICA

To my wife for her love
And her patience and support,
And to my children
For the pride and pleasures that they've brought

Preface

In relative levels of abundance, carbon, nitrogen, and sulfur represent the 13th, 15th, and 18th of the family of elements that make up our planet. These elements represent a level of importance, however, that belies their relatively low position on the scale of abundance. Carbon is the fundamental building block of all the planet's life forms; sulfur is the cement that maintains the integrity and structure of proteins; and nitrogen, in addition to serving as an integral component of the cellular matter of living organisms, is an essential participant in virtually every biological phenomenon in the life process.

As major players in the life-cycling biogeochemical region of the planet called the biosphere, carbon, nitrogen, and sulfur also play roles to varying degrees in the biogeochemical regions encompassed by the biosphere, i.e., the lithosphere (the Earth's crust), the hydrosphere (the oceans, rivers, lakes, and streams), and the atmosphere. Each of these elements cycles among these regions in a kind of dynamic equilibrium along with water, oxygen, phosphorus, and the other elements, in what are defined as grand biogeochemical cycles. But on global, continental, and regional scales, the grand biogeochemical cycles of these three elements are significantly more vulnerable to human interference than are water, oxygen, and phosphorus, all of which are equally critical to life processes. That is, although there may be local disturbances of water balances due to deforestation, urbanization, and the like, or in oxygen balances due to deforestation by fire, or in phosphorus levels due to heavy fertilization, there is little man can do to interfere with their basic global cycles. The planetary cycles of evaporation, precipitation, and runoff of water maintain the Earth's global volume as essentially constant. The very abundance of oxygen, our most abundant element, insulates it from interference. Thus, with its estimated atmospheric mass of about 10^{15} tons, oxygen might suffer a reduction of only about 1%, even if all known deposits of fossil fuels were burned. And phosphorus, having only limited mechanisms for cycling through the biosphere, remains localized even in the event of indiscriminate discharge and, therefore, also lends itself to relatively easy control. In contrast, however, the compounds of carbon, nitrogen, and sulfur, which exist in each of the three states of matter, are not protected by their abundance, and they do move freely among the lithospheric, hydrospheric, and atmospheric biogeochemical regions. Consequently, significant perturbations in their local levels can result in abnormal redistributions of regional, continental, and even global levels.

In terms of environmental impact, therefore, major focus has recently been directed to control of these three life-sustaining elements because of their mobility and their ubiquitous introduction into the biosphere through industrialization and the consumption of vast quantities of energy. And, because they are locked into countless numbers of compounds and generated in countless forms by an enormous wealth of agricultural and industrial processes, need for their monitoring and control reaches into nearly all of civilization's activities. Abnormal levels anywhere in the atmosphere, the lithosphere, or the hydrosphere may have impact on the biosphere. Thus, their sources and cycle pathways must be identified and quantified.

Traditionally, measurement of these elements and their compounds has been the domain of the microorganic analytical chemist. But environmental impact has generated an analytical chemical need that has spread beyond the confines of the specialty laboratory. The quality control and research laboratories of a multiplicity of industries—energy, transportation, fertilizers, pharmaceuticals, chemicals, steel, mining, rubber, and waste and water treatment, to mention a few—have become increasingly involved in monitoring activity. Accordingly, analysis costs and labor intensity have become important factors in selection of methods and acquisition of apparatus; and those charged with selection are, more often than not, untrained in analytical chemistry.

Although there are rich resources available to help the nonspecialist make selections, they are in the form of thousands of individual publications, with each, more often than not, directed to a specific approach and a specific analytical problem. Treatises on organic analysis generally include sections on carbon, nitrogen, and sulfur. However, the information in them is usually presented from a classical analytical chemistry viewpoint and provides minimal reference to environmental applications. These kinds of resources can be intimidating to the untrained technician reponsible for selecting the method appropriate to his problem; and a single compilation that addresses the principles of appropriate analytical methods along with suggestions as to their strengths and weaknesses has not been available.

This monograph, therefore, is directed not to the trained analytical chemist, who has familiarity with and the ability to select methodology from the literature. Rather, it is intended for those engineers and chemists who are suddenly faced with a request to determine the level of sulfur in the factory's stack gas or carbon in its waste stream, as examples, but who have always worked in unrelated work areas. Its goal is to present the underlying principles and some of the limitations of available methods to analyze for carbon, sulfur, or nitrogen—including newer approaches, such as fluorescence, chemiluminescence, and ion chromatography, that are not covered in most of the existing treatises on organic analysis. And with the thought that method development is the province of the analytical chemist, not the environmentalist, and also as a way to avoid a volume of encyclopedic proportions, the discussion, with the exception of the section on analytical applications of enzymes and immunochemicals, has been confined to procedures that are generally available commercially. The material in the monograph is presented at a level believed to require no more background than is provided by typical introductory college level courses in chemistry and calculus.

It is my hope that this monograph can fill a gap between the massive number of journal articles on analysis of these elements and the numerous treatises on analytical organic chemistry, all of which provide excellent method descriptions but do not generally address applications to environmental pollutants.

Jerome C. Greyson

Contents

Part I
Introduction

In today's world we pretty much take our technological prowess for granted. We understand the properties of matter and the behavior of energy. We identify and speak of air and water pollutants, and we give them names like sulfur, nitrogen, or carbon oxides, or we refer to acronyms like TOC, COD, or BOD. Yet we sometimes neglect the fact that our technical arts have been around for millenia, improving in efficiency and accelerating in magnitude as time passed, but all the while generating those things that the names and acronyms represent. The need to monitor and control the profusion of those elements throughout our environment, however, has only recently been recognized. Perhaps it is because it is only recently that population (and its coincident material demands) has grown to a point that has stimulated their generation beyond a level of local irritation to a threat of continental and even global dimensions. You see, among those things that we have named and assigned acronyms to, the waste products of our modern industrial society, are some of the essential elements in the grand biogeochemical cycles of our planet. And they are colliding with the natural order of those cycles. It seems worthwhile, therefore, in a book on the analysis of carbon, nitrogen, and sulfur pollutants, to explore, at least briefly, how it all started, to discuss the origins of these elements as pollutants, and to investigate why they play such important roles in the life cycles of the biosphere.

1

1

Pollutants in the Environment: Past and Present

One might argue that the seeds of environmental pollution were sewn when man lit his first fire, or at least when he smelted his first sulfide ore to recover copper. He was surely familiar with the noxious odor of sulfur dioxide. The history of environmental pollution, however, is not so much a history of fire in all its technological forms, or even a history of technology, as it is a history of population growth. Except for incidents of plague in the Middle Ages, one finds little recognition or even mention of pollution as a societal problem until about the middle of the nineteenth century, when the industrial revolution was well underway, changing society from a predominantly rural to an urban culture. And it was only 50 years or so earlier, with Lavoisier's closing of the age of alchemy, that our key contributors to pollution, carbon, sulfur, and nitrogen, in all their chemical forms, were even recognized as distinct chemical moieties.

PRE-INDUSTRIAL REVOLUTION SANITATION

To early man, fire was survival. He was a hunter-gatherer who used fire to cook his food, keep him warm, and ward off predators. Early man lived in harmony with his environment, adapting to it, and taking his needs from his surroundings as they were presented. Hunting and gathering in one region, however, would not support population growth. At best, the lands of the hunter-gatherers supported about two beings per square mile. So man wandered, following the wild herds that provided his sustenance [Bronowski 1973, 45].

About 10,000 years ago, a change came about in man's cultural pat-

3

terns. He had learned to domesticate animals and to cultivate wild grains; and these new technologies anchored him. Permanent communities were established and food became more plentiful, though dependent upon the capriciousness of the weather. So, by about 3200 B.C., irrigation practices began to help increase and stabilize food production [Drower 1954,1:521]. And, by the middle of the second millenium B.C., aquaducts, reservoirs, even settling ponds and cisterns for domestic water began to be constructed to stabilize water supply as well.

The Assyrians (c. 1000 B.C.) were the masters of this hydraulic technology, constructing water supply systems that rival even some modern engineering projects. The Assyrians were also the first to establish water use regulations, defining the rights and responsibilities of land owners whose properties drew water from communal sources [Carr 1966, 30–35]. Land owners were expected to keep water canals free of silt and pollution and ensure that downstream users got their fair share.

The ancients also practiced a form of sanitation [Carr 1966, 37–39]. As they learned to drain swamps to reclaim land for farming, they applied their technology to the construction of city sewers which served first as storm sewers and secondly to drain sanitary waste. The latter was piped from household privies to the sewers or dumped directly, allowing the rains to carry it off to the local rivers and streams. The systems worked but their sanitary efficacy was somewhat seasonably dependent.

Like the Assyrians, the Romans were also master hydraulic engineers. They constructed aquaducts that transported water from sources located at great distances from Rome. Drinking water was led into settling tanks and cisterns for the city's domestic supply, and undrinkable water from hot springs was used for the city's fountains, to flush sewers, or to drive mills. Despite their marvelous water supply systems, the Romans still practiced sanitation similarly to the Assyrians. Storm sewers served double duty.* But Roman citizens were provided with public baths and public toilets that were flushed by neighboring fountains, and the Romans appeared to be at least aware of the dangers of pollution. The great Roman physician Hippocrates spoke of the need for pure drinking water and wrote essays about the effects of airs, water, and localities (i.e. environment) on life [Cohn and Metzler 1973, 3].

With the rise of Christianity, the magnificent water supply and sewer

---

*Early European and American engineers also constructed storm sewers that eventually were made to serve double duty. As a result, most modern sewage disposal plants must treat storm drainage as well as sanitary sewage. To this day, during heavy rain periods, it is not unusual for a plant to be overloaded and be forced to discharge untreated sewage to its outfalls—either that or let it back up into the city's streets.

technology of the Assyrian and Roman Empires disappeared. By the Middle Ages, cleanliness and sanitation had somehow become theologically suspect. No longer were there drainage systems, storm or sanitary, to carry off the waste of the village populations. Ignorance allowed latrines and cesspools to be located near wells; disposal of garbage and sanitary waste amounted to throwing it into open gutters in unpaved streets; and the excrement of roving domestic animals added to the general mess. All of this open filth, of course, led to the first of civilization's major pollution incidents, the Black Plague of the fourteenth century, an incident that is said to have wiped out as many as three quarters of the population in the years 1348 and 1349 [Carr 1966, 40–43].

Air pollution in the Middle Ages, despite the chemical arts of the time being reasonably well developed, was essentially inconsequential. There were certainly, however, portents of things to come. In the early thirteenth century, coal had begun to be mined, mostly for black-smithing but also for domestic fuel. Its use, however, provoked an outcry from city residents, especially the monied ones, because of the smoke and sulfurous fumes it generated. So charcoal remained the material of choice for most metallurgical activities, and wood the domestic fuel of choice. But consumption of charcoal and wood soon started to deplete forest resources, forcing increased use of coal; and by about 1300, enough smoke was being generated over London to convince parliament to outlaw its use as a heating fuel, a prohibition that seems to have been mostly ignored. Coal continued to be used despite this and other prohibitions, increasing the pall over the city. Interestingly, in 1661 in London, John Evelyn, a prominent scholar of the time, proposed that there was a relationship between the city's smoke pall and disease, and he called for more stringent controls. Evelyn was ignored, of course [Lewis 1965, 9–20]. It took the disasters of later centuries to stimulate some serious action.

THE INDUSTRIAL REVOLUTION: CHOLERA AND KILLER SMOGS

By the end of the eighteenth century, cleanliness and sanitation again became priorities. The industrial revolution was budding, centralizing industry in the cities; and cities were becoming bigger. Streets were cobblestoned and storm drains were installed. The toilet and sink trap had been invented [Carr 1966, 41]; and by the early part of the nineteenth century, water closets discharging to cesspools had become fairly common. As might be expected though, since man is often remiss in his responsibilities; cesspools were only infrequently drained. Especially in the bigger cities, their overflow became a significant source of ground contamination and it soon became necessary to discharge sanitary waste to the storm drains.

Although sewage is mostly water, its residual solids (about 0.1%) contain all manner of organic matter which serves as the nutrient base for all manner of infectious organisms. And before the time of sewage treatment, ground contamination became river contamination, and the rivers were the sources of domestic water. By the middle of the nineteenth century, cholera had killed 20,000 people in London, and countless more had fallen to typhoid and typhus in other parts of the civilized world [Carr 1966, 43].*

Nineteenth century industry and urbanization brought increasing air pollution as well. The Leblanc process for producing sodium carbonate was developed in 1789 and was in full bore by 1825, discharging fumes and smokes of hydrochloric acid and calcium sulfide. And by the middle of the nineteenth century, heavy chemical industry was a common landscape phenomenon in the British Isles and on the European continent—the smoke-blackened skies and soot-marked buildings seen in old photographs bear testimony to the toll that was being taken.

The discovery of microorganisms and the introduction of waste treatment and water sterilization helped to mediate water problems by the 1880s, but air pollution worsened. Despite government regulations,† the booming growth of chemical industry as well as population with its attendant fuel consumption were causing an ever increasing amount of pollution to enter the atmosphere. Oil was discovered in Titusville, Pennsylvania in 1859, and that burgeoning industry, providing a new kind of fuel, as well as a host of daughter products, contributed even more to the pall. And along with the pall, the late nineteenth century brought the first episodes of death-dealing smog.

H.R. Lewis relates an interesting tale [Lewis 1965]. In 1542, the Spanish explorer Juan Cabrillo entered what is now Los Angeles' San Pedro Harbor. In his diary, he wrote not only of the magnificence of the Los Angeles basin with its ring of mountains, but also of the smoke from the Indian fires that rose straight up a few hundred feet and then, as if encountering an invisible barrier, spread horizontally over the valley. He named the harbor the Bay of Smoke.

---

*Untreated sewage to this day is an environmental problem. A goodly percentage of the hospital beds in underdeveloped nations remain devoted to treatment of sewage-related infections. Even in developed nations, sewage outfalls have contaminated estuaries and rivers, making consumption of shell fish a chancy proposition.

†The English Alkali Act of 1863 required that at least 95% of the acid emissions from sodium carbonate plants be retained.

So, temperature inversions are hardly a modern phenomenon; only their consequences are. Probably the most notorious inversion-related pollution problem is that in Los Angeles, where a combination of heavy automobile traffic, heavy industry, meteorology, and terrain maintain a continuing pall of eye-burning haze over the city. Despite the establishment of an Air Pollution Control District in 1947 to combat the problem, only moderate progress has been made to date. But Los Angeles has been fortunate. Its problem, though serious, has never reached the magnitude that some other cities' problems have. Los Angeles has never experienced a killer smog.

In contrast, in December 1873, in London, industrial- and fuel-sourced air pollution killed 1150 people. The same city experienced similar incidents at least four other times between January 1880 and December 1882. And in Glasgow, in 1909, an air pollution episode felled about one thousand people.*

Killer smogs were not confined to the nineteenth century either. London suffered again in 1956 when over a period of only eighteen hours nearly one thousand people died; and then again in 1962 despite an ongoing air pollution control program in the city, another episode killed 750. And nearly the whole of the British Isles was blanketed in toxic smog for a four-day period in December 1952, when some 4000 deaths occurred.

The U.S. has had its share of incidents as well. In October 1948, a combination of fog accompanied by a thermal inversion trapped a high concentration of sulfur dioxide over Donora, Pennsylvania, a small but heavily industrialized city near Pittsburgh. Nearly half the city's population became ill and twenty people died. And in November 1953, in New York City, an eleven-day period of windlessness and thermal inversion allowed an accumulation of pollution from as far away as Ohio and Pennsylvania to invade the city. Some 175 to 260 deaths were ultimately attributed to that incident.

Killer smogs, incidentally, are different in character than automotive smog. They are composed, primarily, of sulfur dioxide and particulates, and they are generated mostly from combustion of unrefined fossil fuels containing high sulfur levels. Automotive smog, on the other hand, is a product of light-induced reactions between exhausted nitrogen oxides and unburned volatile organics. The eye- and chest-irritating products of the reaction are nitrogen dioxide, ozone, and peroxyacetylnitrate. The history

---

*The name "smog" was given to these killing clouds by H. A. Des Voeux in a 1911 report on the Glasgow incident.

of the proliferation of both types of smog, however, is the history of increasing fuel consumption for transportation, to drive industrial processes, to generate electricity, and to heat homes.

DEALING WITH THE PROBLEMS

Although they were certainly tragedies, the cholera epidemics and killer smogs of the nineteenth century, as well as those of more recent origin, have had some positive consequences. They focused attention on the dangers of unconstrained release of waste, and in many cases they stimulated corrective action, albeit, frequently just moving the problem to someone else's backyard.

London, for example, of all the major European cities, had suffered the most damage from cholera, and recognizing that the source was indeed sewage-contaminated water, constructed sewers that carried their waste well downstream of the city. This resolved the city's immediate problem, but one wonders about those poor souls living near the downstream outfall.

By 1880, however, Karl Joseph Eberth had discovered the typhoid bacillus; and in 1883, Robert Koch isolated the cholera microorganism. Pasteur's discovery of sterilization in combination with the invention of sand filtration started the road to modern water treatment and to routine microbiological examination of water supplies.

Sewage treatment also started in the late nineteenth century, one of the first treatment plants in the U.S. was built in Memphis, Tennessee in 1880 [Environmental Quality 1981, 51]. Early sewage treatment, however, was usually limited to filtering and sedimenting gross solids from the influent, with the clear effluent being returned to a neighboring waterway, sometimes sterilized, sometimes counting on the receiving water's dilution capability, but always carrying high concentrations of organic matter. The danger of proliferating infectious organisms with this "primary treatment" was reduced. But the high organic burden returned to the receiving water became a major contributor to erosion of its quality by reducing its dissolved oxygen levels, making it anaerobic, killing its marine life-forms, and then generating some pretty toxic and obnoxious products. By 1927, a national survey in the U.S. [Hatfield 1927] indicated that more than 85% of the country's inland waterways were polluted. This led to increased federal action in the 1930s and construction of a large number of new municipal treatment plants. Many of these incorporated secondary processes that could reduce the level of organic discharge, but could not eliminate nitrates and phosphates, which continued to be returned to the

receiving water, effectively fertilizing it, stimulating growth and proliferation of large algal blooms, and accelerating its eutrophication.

In contrast to water pollution which received attention because it was
a clear threat to public welfare, and despite killer smogs, early air pollution
seemed to have been viewed by the powers-that-be as more a nuisance
than a public health problem. It was really not until the forties and fifties
of this century that it began to be generally regarded as serious. Early
smoke control legislation addressed immediate and local problems,* so
gas scrubbers were installed and stack heights were raised to improve
smoke dispersion and dilution. But politics and economics complicated
enforcement of early air pollution legislation. Local and state officials
were frequently reluctant to impose or enforce really tough emission
standards, fearing that industry and its associated tax base might move to
more accommodating communities. Furthermore, even when they were
enforced, like primary sewage treatment, those early smoke control regulations frequently just changed one kind of a problem for another. High
smokestacks moved the toxic gases from their immediate sources into the
upper atmosphere where they could engage in another kind of mischief.

The force that probably broke the regulatory camel's back and started
serious examination of air pollution, was not stack gas or killer smog, but
automotive exhaust. After the end of World War II, the population
of automobiles in the U.S. increased enormously. The increase was
accompanied by equally large increases in airborn concentrations of carbon monoxide, nitrogen oxides, and smog. The problem was further
exacerbated by the retirement of electrically-powered light rail public
transportation systems (trolley cars), which automobile, petroleum, and
tire interests heartily encouraged,† and which were replaced by buses and
more cars. In Los Angeles, the resulting constant eye-burning, cough-
inducing haze from combined vehicular and industrial emissions forced
the creation of the Los Angeles Air Pollution Control District. It also
focused national attention on the severity of damage environmental pollution could inflict.

But still, not a great deal of attention was paid to the "environment"
at the national level until the seventies. The sixties brought increasing
attention to the problems, and the Federal Government did create several

---

*By 1912, 23 of the 28 largest cities in the U.S. had enacted smoke control legislation
[Environmental Quality 1981, 21].
†One major automobile manufacturer in concert with a major petroleum company and a tire
manufacturer bought up the entire light rail systems of a number of cities, including Los
Angeles. They immediately retired the systems, tore up the tracks, and replaced the trolleys
with diesel- and gas-powered buses.

new research agencies to study pollution, but there was a growing sense of frustration in the public about the level of effort being expended. Also, in the early and middle sixties, a series of books appeared in the popular press [Carson 1962, Carr 1966, Battan 1966] that detailed the consequences of continuing environmental neglect, and intensified the perception that federal activity had no teeth, and that pollution problems had grown well beyond local or state resolve (or ability) to address. The intensity of public opinion culminated in an "Earth Day" demonstration in April 1970 in which thousands of people called for national action to control pollution.

As a result, Congress combined the individual research agencies into one oversight organization by establishing the Environmental Protection Agency. It also passed the Clean Air and Clean Water Acts, which set national air and water quality standards and defined rules and limits for pollutant discharges.

Emission limits were defined for "criteria air pollutants," which were those found most commonly throughout the country and included carbon monoxide, sulfur oxides, and nitrogen oxides, as well as particulates, lead, and ozone. Water quality standards were established that were based on receiving water's normal use. Thus, discharge into recreational waters would be more restrictive than discharge into navigable waters. Dischargers were thereafter required to obtain permits specifying the type and amount of pollutants they could release; they were obliged to maintain monitoring programs and report discharge data to EPA; and they became subject to compliance inspections. EPA's list of substances categorized as pollutants included heat, organic wastes, sediments, acid, bacteria and viruses, oil and grease, toxic substances like heavy metals, pesticides, and solvents, and phosphates and nitrates.

The impact of this new legislation was rapid. A 1983 survey of the nation's waterways indicated that of the 354,000 miles of rivers and streams that were monitored, 47000 miles showed quality improvement, 11000 miles became worse, and 296,000 miles were about the same [Environmental Progress & Challenges 1984, 46]. As for the "criteria air pollutants," by 1980, dramatic decreases in emissions of particulates, carbon monoxide, and volatile organics had been achieved along with smaller but significant reductions in sulfur oxides [Environmental Progress & Challenges 1984, 15].

BEYOND SMOG AND EUTROPHICATION

Thus, the establishment of standards and the assiduous enforcement of controls in the seventies and eighties achieved reductions in levels of the more obvious pollutants and gave promise for continued improvement. But during the same period, pollution research activity began to focus on issues that extended beyond the immediate effects manifested as smog or local waterway pollution. These issues included the erosion of ground water quality, the impact of acid rain, the depletion of the upper atmosphere's ozone layer, and the potential for climate modification due to greenhouse warming.

One of the major contributors to the erosion of ground water quality has been one of our subject elements, nitrogen, in the form of nitrates leached from deposits of inorganic fertilizers and animal wastes by rain and irrigation water. With the introduction of high production technology into agriculture, applications of inorganic fertilizers have increased enormously. For example, the annual consumption of nitrate fertilizers exceeded 10 million tons in 1980 [Smil 1985, 158]. And even more nitrogen may be entering underground aquifers as a consequence of a U.S. population of about 160 million large-farm and ranch animals [Nyberg 1987, 1-G] that freely deposit wastes on the ground equivalent to a human population of 330 million. The net result of all this is dramatized by the amount of nitrogen burden entering surface waters in only one agricultural state, Wisconsin. There, in the late sixties, surface waters were absorbing about 80 million pounds of nitrogen per year of which over 40% was attributed to percolation of contaminated ground water [Hasler 1970, 112].

Interestingly, another statistic from Wisconsin's surface water nitrogen burden illustrates how complex environmental pollution problems can become, especially with respect to carbon, nitrogen, and sulfur. About nine percent of the nitrogen entering Wisconsin surface waters was attributable to nitrates carried there by rain and snowfall, a demonstration of the mobility of nitrogen as an environmental pollutant. It can move not only as a soluble component of water, but as a component of air, with about 7 million pounds per year of airborn nitrates contributing to the nitrogen burden of Wisconsin's rivers and streams. Mobility is characteristic of carbon and sulfur pollutants as well. It is the reason for their dispersion over regional, continental, and even global dimensions in the form of acid rain, and for their contributions to the depletion of stratospheric ozone levels and to greenhouse warming.

Damage from acid rain, incidentally, was recognized as early as the late nineteenth century [Acid Rain 1986]. But it has been only since the late seventies of our century that national attention began to be focused on the general subject of acid deposition, i.e., the rain, snow, clouds, and fog that contain high concentrations of nitric and sulfuric acids, and damage lakes, forests, and building materials remote from possible sources. Studies since then have resulted in recognition that acid deposition is a pollution problem of nearly continental dimensions, seriously threatening forests and lakes in Central Europe as well as the entire east coast of the U.S.

Acid deposition is directly related to emissions of nitrogen and sulfur oxides from automotive exhausts and industrial processes that use coal or high-sulfur oils as fuels. Despite EPA's efforts to control the "criteria air pollutants," in 1980 in the U.S., 21 million tons of nitrogen oxides were thus released to the atmosphere along with 27 million tons of sulfur dioxide [Acid Rain & Transported Air Pollutants 1984, 5]. Though the gases are not themselves acids, it has been recognized that they become so as a consequence of somewhat complicated photochemical reactions with water, ozone, and certain organic oxidants found in the upper atmosphere.

Photochemical reactions are also responsible for the depletion of the ozone layer, sizable fluctuations in which have been observed by atmospheric scientists over the past several years. Ozone levels over the Antarctic in particular have recently been observed to effectively reduce to zero for periods of one or two months a year [Zurer 1987, 25]. Since ozone is unique in its ability to shield the surface of the earth from harmful solar ultraviolet radiation, changes in its upper atmosphere levels are viewed with some alarm.

The processes by which upper atmosphere ozone can be destroyed involve a number of gases, but foremost are nitrous oxide and a widely used group of organic compounds called chlorofluorocarbons. The latter are currently recognized as the major culprits, but the former is a potentially serious contender. Controlling the release of the chlorofluorcarbons requires only that regulatory agencies so order. Control of nitrous oxide, however, is a more difficult problem with which to contend.

With the exception of nitrogen gas itself, nitrous oxide is the most abundant atmospheric nitrogen compound. It is both naturally occurring and anthropogenically generated (man-made). It is an intermediate product in the biological denitrification of nitrates,* and it is a product of the combustion of fossil fuels and biomass materials like wood and grass. So, although it is naturally generated in soil and in the ocean as nitrate-containing plant matter decays, its atmospheric burden is also being elevated by the increased use of inorganic fertilizers as well as the increased burning associated with accelerated industrial, domestic, and agricultural activities. Of the 110 million tons of N_2O estimated to be released to the atmosphere annually, 20% is from denitrification of fertilizer and 14% from fuel and biomass combustion [Smil 1985, 200].

---

*Denitrification is the process by which nitrates in soils and plant matter are reduced to nitrogen gas and returned to the atmosphere.

Nitrous oxide and ozone, along with our other subject pollutants, are also among the so-called greenhouse gases. But except for carbon dioxide, recognition of the significance of these gases as contributors to greenhouse warming is only relatively recent. Greenhouse warming occurs, of course, because after the sun's broad spectrum solar radiation penetrates the earth's atmosphere, some of it is re-radiated back to space in the form of heat, where it is absorbed by upper atmosphere infrared absorbing gases. Thus, the amount of heat that can escape to space is mediated along with earth's surface temperature. The primary mediating gas is carbon dioxide, but ozone, nitrous oxide, methane, sulfur dioxide, nitric acid, and the chlorofluorocarbons are now known to contribute as well [Smil 1985, 92], and they add to the difficulty of quantitatively analyzing and controlling greenhouse warming. For example, although ozone depletion is a potentially serious problem of one kind, it may contribute in a positive way by countering the increased warming that increasing levels of carbon dioxide may cause. On the other hand, the other mediating gases, all of which are air pollutants, may be adding to the warming effect.

The influence of carbon dioxide on the greenhouse effect, incidentally, has been a long-time subject of considerable debate. Though CO_2, as a product of virtually all of nature's processes, is not generally regarded as an air pollutant, fossil fuel combustion has released enormous additional quantities of the gas to the atmosphere, with estimates of upwards of 180 billion tons being released since the beginning of the industrial revolution [Smil 1985, 43]. In that sense, CO_2 is a pollutant; and it is argued that its enormous releases have led to a rise of about 1°K in the Earth's average temperature.

The debate, however, has centered on the great buffering capacity of the ocean as a sink for carbon dioxide and whether or not there has been a real net increase in atmospheric CO_2. Recently, though, atmospheric scientists have recognized that gas exchanges between the atmosphere and the ocean occur only in the surface layers, which comprise only two percent of the ocean's volume, and exchanges between the surface and the deeper ocean layers are too slow to keep up with the rate at which carbon dioxide has been released [Smil 1985, 48]. Thus, the atmospheric concentration of CO_2 has been increasing. And, in fact, measurements made during the period 1959 to 1979 indicated increases in levels at Mauna Loa, Hawaii and in Antartica [Environmental Quality 1981, 254]. Estimates of global levels suggest that values have risen from about 280 ppm (volume) to about 350 ppm since the beginning of the industrial revolution; and if fuel consumption continues to accelerate at its current rate, projections into the next few generations indicate that the CO_2 atmospheric concentration will be twice its present level. Such an increase could cause the average temperature of the globe to rise about 3°C as well as cause polar melting and a rise in sea level.

But increasing attention is now also being directed to the other greenhouse gases since, despite attempts to control their emissions, they too are increasing in concentration in the upper atmosphere. Increases in their levels are now considered to be contributing to surface warming by an amount nearly equal to that of carbon dioxide [Environmental Quality 1985, 212], and concern with what was formerly a "CO_2/climate" problem is now concern with what is regarded as a "trace gas/climate" problem.

CONCLUDING REMARKS

As one concludes this chapter, one should consider the overall complexity implied by the discussion, noting in particular the interrelationships of environmental problems and the second order effects of control systems, as well as the intimate involvement of our subject elements. Carbon, nitrogen, and sulfur have nearly always played some role in pollution, beginning with the noxious airs of the Middle Ages and becoming progressively more important as industrial and agricultural techniques have become more sophisticated. But their roles are now extended beyond the local irritations of smogs or lake contamination into regional and even global problems like acid rain, ozone depletion, and greenhouse warming. And they have assumed these important roles because of their ubiquity in technology and agricultural and their ability to move freely among the various regions of the biosphere.

2

The Biosphere and Biogeochemical Cycles

The biosphere is defined as that part of the planet that sustains life. It encompasses the lower part of the atmosphere, the hydrosphere (the oceans, lakes, rivers, and streams), and the lithosphere (the earth's crust) down to a depth of about two kilometers. Biogeochemical cycles are the transport pathways and the chemical and physical interactions of the elements within and among these regions of the biosphere.

Figure 2.1 is a simplified diagram illustrating some of the modes of material transport among the biospheric regions. Within each of the regions, of course, natural and anthropogenic biological and chemical processes are always underway, producing or consuming the substances that circulate among the regions via the transport modes shown.

THE CARBON CYCLE

Global biogeochemical cycles for individual elements are usually illustrated in schematic diagrams like that of Figure 2.2, which shows the sources, sinks, processes, and fluxes of carbon among the regions of the biosphere. In Figure 2.2, carbon reserves and estimates of their quantities are shown in rectangles. Conversion mechanisms and quantities transported annually within or to other regions are illustrated with the engineer's symbol for a valve.

As can be seen, the major modes of conversion and transport of

---

The statistical data in this chapter have been liberally extracted from Smil's exellent volume [Smil 1985].

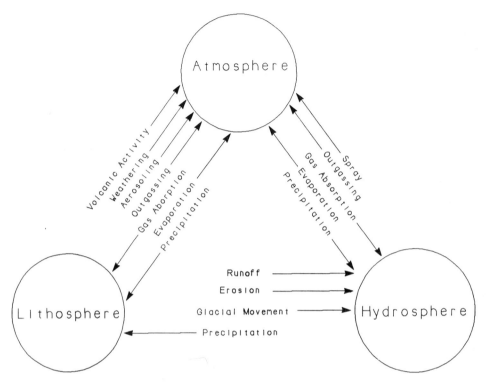

Figure 2.1   Material transport modes among the regions of the biosphere.

carbon include exchange equilibria between atmospheric carbon and soluble carbonates in the hydrosphere, photosynthesis, plant and animal respiration, combustion, and plant and animal death and decomposition. Equations describing these chemical processes are shown in Table 2.1.

Atmospheric carbon represents the smallest of the planet's reserves, containing only 720 billion tons, mostly as carbon dioxide. This quantity is to be compared to the largest of the planet's reserves, about 27 million

Table 2.1   Some Important Reactions of Carbon in the Biosphere

Carbon-Carbonate Equilibrium
$$CO_2 + H_2O = H^+ + HCO_3^- = 2H^+ + CO_3^{2-}$$

Photosynthesis
$$CO_2 + H_2O + light = plant\ carbon + O_2$$

Respiration, Decomposition, and Combustion
$$plant\ carbon + O_2 = CO_2 + H_2O + heat$$

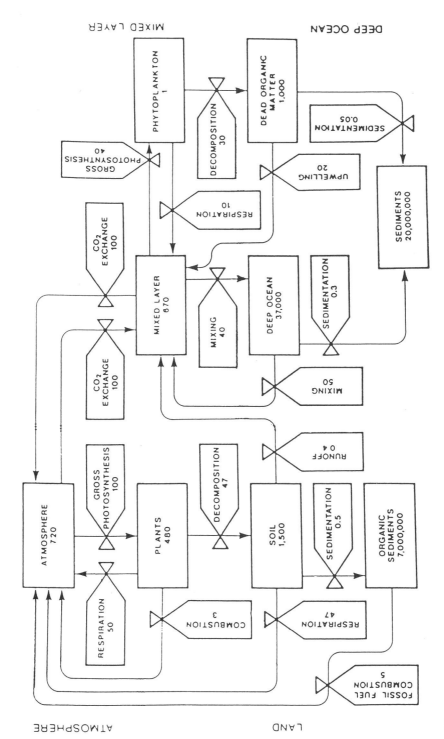

Figure 2.2 The biogeochemical carbon cycle. Storage values and fluxes are in billions of tons. [Smil 1985, courtesy Plenum Press].

billion tons, which reside in the lithosphere and in the deep ocean, mostly as organic sediments or carbonate rock. The hydrosphere is the second largest reserve, containing about 38 thousand billion tons in the form of soluble carbonates and another 1000 billion tons in the form of dead organic matter.

The atmosphere, containing the smallest carbon reserve, is the most vulnerable to fluctuations in carbon transport among the biospheric regions, especially so, since the quantities transported among the reserves are of about the same magnitude as the atmospheric reserve.

Exchange of carbon between the atmosphere and the hydrosphere amounts to about 100 billion tons per year. Roughly 40% of that is consumed in the photosynthesis of phytoplankton, and it is ultimately returned to the atmosphere through respiration and the death and decomposition of the aquatic plants. It is clear that any pollution threat to the ocean's plankton population could portend a disaster for the carbon cycle equilibrium.

Photosynthesis is the primary mechanism maintaining carbon equilibrium between the lithosphere and the atmosphere, producing about 100 billion tons of terrestrial plant carbon per year. Return is through respiration of plants and oxidation of decomposed plant matter in soil. Current plant carbon reserves are estimated to be between 480 billion and a trillion tons, so plant reserves are critical to maintaining the cycle equilibrium. It is noteworthy, therefore, that gross deforestation to produce arable or pasture lands, especially of tropical rain forests, which is one of man's more recent active pursuits, may be threatening plant carbon reserves. Reduction of those reserves is now estimated to be one to two billion tons per year. Continued deforestation could ultimately increase the atmospheric carbon burden by reducing the photosynthetic consumption of carbon dioxide and add to greenhouse warming.

But the major impact of man's activities, at least for the present, appears to be plant and fossil fuel combustion which, as can be seen in Figure 2.2, contributes about eight percent of the total amount of carbon returned to the atmosphere. Plant combustion, representing about 3 billion tons of that, is a form of respiration and can be regarded as part of the photosynthetic cycle; but it still may generate an apparent excess of CO_2, because it is a kind of accelerated respiration. Fossil fuel combustion, on the other hand, definitely adds an excess of about 5 billion tons of carbon to the atmosphere annually.

THE NITROGEN CYCLE

Despite enormous stores of nitrogen in the lithosphere (about 5×10^{16} tons), only a tiny fraction of the total (about 240 billion tons) that resides in soil takes part in nitrogen's biogeochemical cycle. Most lithospheric nitrogen is stored within igneous rock, inaccessible to any mechanism that might incorporate it into the cycle.

The hydrosphere contains the lowest planetary store, estimated to be about 2.3×10^{13} tons, with 95% stored as molecular nitrogen and the other five percent distributed in about a 60:40 ratio between inorganic nitrates and organic nitrogen. But hydrospheric nitrogen plays only a small role in the global cycle; because the major fluxes of nitrogen are related to consumption by plant matter, and the terrestrial plant mass is orders of magnitude greater than the marine mass.

So Figure 2.3, illustrating the nitrogen cycle, deals only with the most dynamic fluxes and the major sources of cycling nitrogen, concerning itself primarily with interactions among soils, the atmosphere, and anthropogenic activities. Some important reactions of nitrogen underway within the cycle are shown in Table 2.2.

The atmosphere contains about 3.5×10^{15} tons of nitrogen, primarily in molecular form but including nitric oxide, nitrogen dioxide, and nitrous oxide. It is the principal source of fixed nitrogen, which is a generic term for nitrogen compounds that are produced by conversion of atmospheric N_2, and which may ultimately serve as plant nutrients. Nitrogen may be fixed biologically, by lightening, by combustion of fossil fuels and biomass materials, and by synthesis of nitrogen fertilizers (Haber process).

Biological fixation of nitrogen in plant matter, forming amino acids

Table 2.2   Some Reactions of Nitrogen in the Biosphere

Natural Nitrogen Fixation

N_2 + starch + bacterial adenosine triphosphate = NH_3

$2N_2 + 6H_2O$ + lightning = $4NH_3 + 3O_2$

Industrial Fixation (Haber Process For Synthesizing Ammonia)

$C + H_2O(steam) = CO + H_2$

$N_2 + 3H_2 = 2NH_3$

Amino Acid Synthesis (Plant Protein)

$2NH_3 + 2H_2O + 4CO_2 = 2CH_2NH_2COOH^{(glycine)} + 3O_2$

Nitrification

$2N_4^+ + 3O_2 = 2NO_2 + 4H^+ + 2H_2O$

$2NO_2^- + O_2 = 2NO_3^-$

Denitrification

$6NO_3^- + glucose = 6CO_2 + 3H_2O + 6OH^- + 3N_2O$

$4NO_3^- + 2H_2O = 2N_2 + 5O_2 + 4OH^-$

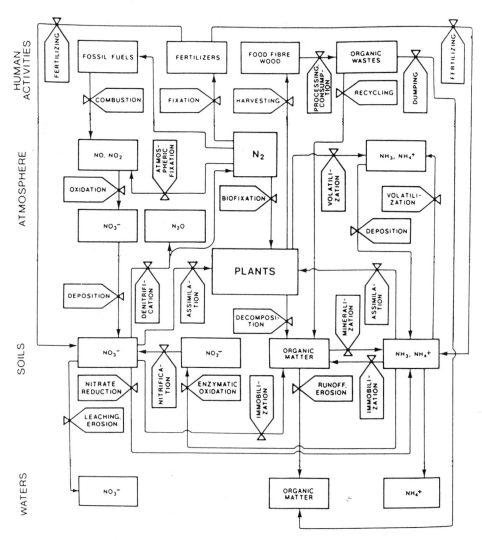

Figure 2.3 Nitrogen's biogeochemical cycle [Smil 1985, courtesy Plenum Press].

and plant proteins, occurs as a result of symbiotic relationships between plants and bacteria, like those between *rhizobium* and soybeans or alfalfa, but also including numerous tropical and subtropical trees. Nitrogen may also be fixed in soil by bacteria like *clostridium* and *azotobacter* which contain the enzyme nitrogenase, a catalyst for the reduction of atmospheric nitrogen to ammonia. The ammonia is then assimilated by plants in another path to protein synthesis.

Biofixation represents the most important supply of nitrogen to plants and to the soil, contributing 130–150 million tons per year. Synthesis of nitrogenous fertilizers from atmospheric nitrogen is not far behind, however, contributing a third to half the contribution of biofixation. Lightening and combustion of fossil fuel and biomass also fix atmospheric nitrogen, converting it to nitrogen oxides. Lightening fixes about 10 million tons annually. The amount of atmospheric nitrogen fixed by combustion, however, is difficult to distinguish from the nitrogen oxides formed by the oxidation of the nitrogen intrinsic to fossil fuels and biomass.

Plant nitrogen may cycle within a food chain or it may be returned to the soil as a consequence of death and decomposition. If it enters the food chain, it will ultimately return to the soil as animal waste. Order of magnitude estimates of the quantity of nitrogen thus returned globally are as high as 75 million tons, a quantity even larger than that of inorganic fertilizer deposition. Though animal waste deposits may appear to be simply part of the equilibrium cycle, they represent a conversion of a more or less immobile and dispersed form of nitrogen, i.e., plant protein, to a form similar to inorganic fertilizers, one that is relatively localized, mobile, and accessible to runoff and leaching.

Return of soil nitrogen to the atmosphere results from two processes, the major one being bacterial denitrification of decomposed organic nitrates and inorganic fertilizers. Biomass combustion is the second return path, converting organic nitrogen to nitrogen oxides, and is the second largest anthropogenic contribution of nitrogen to the nitrogen cycle, exceeded only by the deposition of nitrogenous fertilizers.

While combustion is an oxidation process, bacterial denitrification is a chemical reduction process, resulting in nitrous oxide and molecular nitrogen. Denitrification bacteria are numerous, the genus *Pseudomonas* being among the most common. All, however, carry a class of enzymes called reductases, which catalyze the reduction of nitrates to nitrites, then to nitrous oxide, and ultimately to molecular nitrogen. Estimates of the amount of soil nitrate returned to the atmosphere as a result of denitrifying bacteria are of the same magnitude as the amount fixed by plant matter and nitrogen fixing microorganisms. It is a consequence of denitrifying processes, as well as the relatively inert character of nitrous oxide in the lower atmosphere, that makes that gas the second most abundant nitrogen compound in the atmosphere and an increasing threat to the ozone layer, as depositions of inorganic fertilizers increase.

THE SULFUR CYCLE

The biogeochemical sulfur cycle is shown in Figure 2.4. As can be seen, the largest global reserve of sulfur resides in the lithosphere, which contains 10,500 billion tons and which is distributed among igneous and

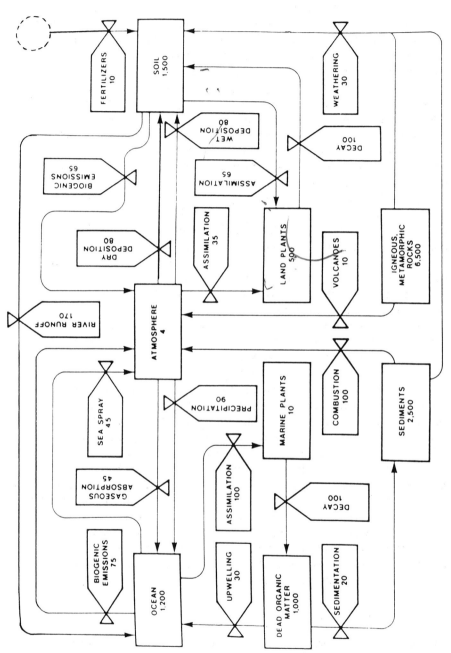

Figure 2.4   The sulfur cycle. Values are in billions of tons of sulfur. [Smil 1985, courtesy Plenum Press].

metamorphic rock, sediments, and in soils. The ocean is the second largest reserve, containing about 2210 billion tons, 1200 billion tons of which is the form of soluble sulfates, with the remainder distributed between living and dead organic matter. Sulfate is the second most abundant anion in sea water, second only to chloride.

The atmosphere, the last of the reserves, is three orders of magnitude lower in sulfur concentration than the oceans or soils, despite the large flows into it indicated in Figure 2.4. The primary components in the atmosphere are hydrogen sulfide, sulfur dioxide, sulfates, and dimethyl sulfide. Biogeochemical equilibrium is maintained by relatively short atmospheric residence times as well as large return flows via absorption and precipitation into both land and the oceans. Residence times for sulfur compounds in the atmosphere have been estimated to range, in the extremes, between several hours and several weeks, short enough to keep atmospheric levels low but long enough to allow sulfur compounds to be carried long distances away from sources, as in acid rain.

Some environmental chemical modifications of sulfur in the biogeochemical cycle are shown in Table 2.3. As can be seen, environmental sulfur may be subject to modification through a number of mechanisms including bacterial action, weathering, combustion, and assimilation by plant matter. Anaerobic bacterial decomposition of organic sulfur is the largest natural source of atmospheric sulfur, amounting to 140 billion tons annually from the oceans and soils in the form of hydrogen sulfide and dimethyl sulfide.

Other major atmospheric entry routes include volcanic activity, sea spray, and fossil fuel combustion. Volcanic activity is the smallest of the contributors (10 billion tons annually) but, in effect, may be more signifi-

Table 2.3   Sulfur Conversions in the Biosphere

Aerobic Bacterial Reactions
$$H_2S + O_2 + (e.g., \text{ thiobacillus}) = S + H_2O = SO_4^{2-}$$
$$\text{organic sulfur} = H_2S = S = SO_4^{2-}$$

Anaerobic Bacterial Reactions
$$\text{organic sulfur} + (e.g., \text{ desulphovibrio}) = H_2S$$

Oxidative Weathering
$$\text{organic sulfur, free sulfur, or metallic sulfides} + O_2 = SO_4^{2-}$$

Combustion
$$\text{organic sulfur or sulfides} + O_2 = SO_2$$

Plant Assimilation
$$SO_2 \text{ or } SO_4^{2-} + \text{plant matter} = \text{organic sulfur}$$

cant than sea spray (45 billion tons). It is thought that about 90% of sea spray sulfur is returned directly to the oceans by absorption and precipitation, with the remainder being deposited primarily over coastal regions.

Combustion of fossil fuels, primarily coal, is the largest anthropogenic contribution, representing between a third and a half of the total of the natural sources of atmospheric sulfur. But perhaps more significant than the absolute amount of sulfur transferred to the atmosphere globally is the fact that the bulk of the anthropogenic emissions are concentrated in the industrialized northern hemisphere, where they exceed natural emissions many fold and can serve as the sources of killer smogs and acid rains.

CONCLUSION

Clearly, the foregoing discussion has been, at best, a second or third order abstraction of the current state of knowledge of biogeochemical cycles. There is a great deal of data available. However, despite the large store of data, the complexity of the cycles is such that there is only limited understanding of the interactions and interferences underway between natural and anthropogenic transfers. Continued measurement of the levels of these major circulating elements in the environment serves a dual role therefore. Not only do such measurements help to constrain the proliferation of pollutants, they ultimately should provide enough additional information to be able to establish long term controls and to manage the future.

Part II
The Chemistry of Carbon, Nitrogen, and Sulfur as Pollutants

3
Carbon, Nitrogen, and Sulfur Chemistry

As discussed earlier, carbon, nitrogen, and sulfur pass among the regions of the biosphere in the form of small simple molecules, primarily oxides, sometimes hydrides. Carbon dioxide, nitrogen dioxide, and sulfur dioxide are representative of the former while ammonia, hydrogen sulfide, and methane are representative of the latter. Environmental monitoring, consequently, is directed primarily toward these simple species except, perhaps, when specific toxic chemicals must be identified for regulatory purposes. We shall, therefore, confine our discussion of the chemistry of our subject elements to the formation and reactions of these small simple molecules, although we will, later on, discuss analytical methods that may be used to identify and quantitate specific and more complex chemicals as well.

Carbon, nitrogen, and sulfur share an important common quality—each may assume multiple oxidation states. That is, each can exhibit multiple valences, behaving as an electron donor or as an electron acceptor. The compounds representing the multiple oxidation states share, additionally, still another common quality: the oxides and hydrides are all gases, providing our subject elements with the mobility to circulate freely among the regions of the biosphere. Multiple oxidation states are, of course, not unique to these elements. What is unique about them, in contrast to others that may exhibit similar behavior, is the role they play in living organisms and their ubiquity in natural and anthropogenic processes.

THE ELECTRON CONFIGURATION OF THE ELEMENTS

The ability to exhibit multiple oxidation states is, of course, a consequence of the electron configuration about the nucleus of an atom. The number of electrons resident in that configuration constitutes the oxidation state of the element. Thus, an atom with a number of electrons equal to its nuclear charge is neutral while one with fewer electrons than the nuclear charge is said to be oxidized and one with a greater number is said to be reduced.

One may think of the electron configuration as a series of shells of increasing diameter, each of which contains a group of electrons rotating about the nucleus in circular, elliptical, and other more complex paths paths called orbitals. The number of electrons in the shells surrounding the nucleus balances the nuclear positive charge and is, therefore, equal to the atomic number of the element and follows the sequence of the Periodic Chart of the Elements (Figure 3.1).

The number of shells surrounding the nucleus is determined by the number of inert gas configurations that the atom's electron population will allow. A filled electron shell constitutes the configuration of an inert gas. Thus, the lowest shell, i.e., shell No. 1 is full when it contains two electrons, since it then has the configuration of helium. As the atomic number increases beyond two, shell No. 2 fills until the electron configur-

IA	IIA	IIIB	IVB	VB	VIB	VIIB	VIII	VIII	VIII	IB	IIB	IIIA	IVA	VA	VIA	VIIA	Zero
1 H 1.008																1 H 1.008	2 He 4.003
3 Li 6.940	4 Be 9.013											5 B 10.82	6 C 12.011	7 N 14.008	8 O 16.000	9 F 19.00	10 Ne 20.183
11 Na 22.991	12 Mg 24.32											13 Al 26.98	14 Si 28.09	15 P 30.975	16 S 32.066	17 Cl 35.457	18 Ar 39.944
19 K 39.100	20 Ca 40.08	21 Sc 44.96	22 Ti 47.90	23 V 50.95	24 Cr 52.01	25 Mn 54.94	26 Fe 55.85	27 Co 58.94	28 Ni 58.71	29 Cu 63.54	30 Zn 65.38	31 Ga 69.72	32 Ge 72.60	33 As 74.91	34 Se 78.96	35 Br 79.916	36 Kr 83.80
37 Rb 85.48	38 Sr 87.63	39 Y 88.92	40 Zr 91.22	41 Nb 92.91	42 Mo 95.95	43 Tc [99]	44 Ru 101.1	45 Rh 102.91	46 Pd 106.4	47 Ag 107.88	48 Cd 112.41	49 In 114.82	50 Sn 118.70	51 Sb 121.76	52 Te 127.61	53 I 126.91	54 Xe 131.30
55 Cs 132.91	56 Ba 137.36	57 *La 138.92	72 Hf 178.50	73 Ta 180.95	74 W 183.86	75 Re 186.22	76 Os 190.2	77 Ir 192.2	78 Pt 195.09	79 Au 197.0	80 Hg 200.61	81 Tl 204.39	82 Pb 207.21	83 Bi 209.00	84 Po 210	85 At [210]	86 Rn 222
87 Fr [223]	88 Ra 226.05	89 ♯Ac 227															

*LANTHANIDE SERIES	58 Ce 140.13	59 Pr 140.92	60 Nd 144.27	61 Pm [145]	62 Sm 150.35	63 Eu 152.0	64 Gd 157.26	65 Tb 158.93	66 Dy 162.51	67 Ho 164.94	68 Er 167.27	69 Tm 168.94	70 Yb 173.04	71 Lu 174.99
♯ACTINIDE SERIES	90 Th 232.05	91 Pa 231	92 U 238.07	93 Np [237]	94 Pu [244]	95 Am [243]	96 Cm [248]	97 Bk [247]	98 Cf [251]	99 Es [254]	100 Fm [253]	101 Md [256]	102 No [253]	

Figure 3.1 The Periodic Chart of the Elements. Note the inert gases in the rightmost column, the outer electron orbitals of which are always full.

ation of neon is achieved. Then shell No. 3 fills until argon's electron configuration is achieved, and so forth up to radon. The highest shell, No. 7, fills after the radon configuration is achieved and starts the sequence of the heaviest elements.

The orbitals within each shell are designated s, p, d, and f. There is one s orbital, and there are three p orbitals, five d orbitals, and seven f orbitals. Just as filling the shells is progressive with atomic number, filling the orbitals within the shells is progressive. In the first three shells, the filling order is first s then p. The order in shells 4 and 5 is s, $(n - 1)d$, and p where n is the shell number; and the order in the heaviest elements is s, $(n - 2)f$, $(n - 1)d$, and p.* The outermost electrons are the valence electrons that take part in bonding to other atoms.

The order of orbital entry is also the order of the resident electron energy. That is, each of the shells and their associated orbitals may also be regarded as energy levels, with electrons resident in the lowest levels having the lowest energies. Thus, promotion of an electron from a lower to a higher level, when that level is vacant and can accept the electron, requires energy input.

Each orbital can hold a pair of electrons.⁺ Thus, each s orbital can contain two electrons while the three p orbitals, sometimes designated p_x, p_y, and p_z can hold a total of six. In the same way, d orbitals may contain 10 electrons and f orbitals may contain 14.

A convenient way to illustrate the electron configurations of the elements is shown in the atomic orbital diagrams of Figure 3.2. As can be seen, helium has the atomic number 2 and two electrons fill its 1s orbital. Correspondingly, neon has the atomic number 10 and is shown with its 1s, 2s and $2p_{x,y,z}$ orbitals filled with a total of 10 electrons. Note that except for helium, the outer shell of which may contain only two electrons, the outer shells of the other inert gases always contain eight electrons resident in the s and p orbitals.

Iron's electron configuration is shown to illustrate the filling order of the heavier elements. Iron's 3d orbitals fill before its 4p orbitals, which are shown as vacant. A point of some consequence to later discussions is that electrons may be excited from filled to vacant orbitals. So, in theory, iron's 3d electrons could be promoted to 4p orbitals if they ab-

---

*The number of orbitals and the order of entry have been determined from theoretical analysis of the emission spectra of the elements.

⁺Associated with every electron is a magnetic dipole, considered to be generated by the electron's spinning about its own axis. Quantum mechanical considerations dictate that only electrons of opposite magnetic polarity may pair to occupy a single orbital.

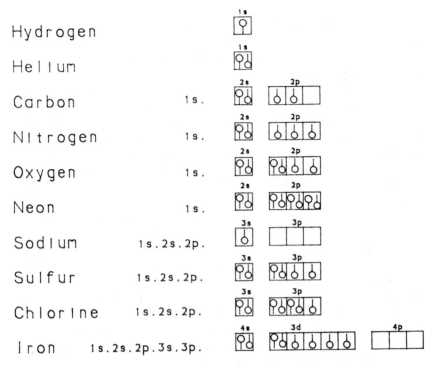

Figure 3.2   Atomic orbital representations of some selected elements.

sorbed the proper amount of energy. The theory of exciting electrons to higher level orbitals and associated higher energy states belongs to the field of quantum mechanics and is beyond the scope of this book. Suffice to say here that the energies absorbed and radiated by the atoms as a consequence of the movement of electrons between orbitals of different energy levels provides the basis of several areas of analytical spectroscopy, and is a topic that will be discussed in more detail later.

CHEMICAL BONDING

There are some basic rules involved in the filling order of the orbitals that bear on both oxidation states and the chemical binding characteristics of the elements to themselves and one another. First, referring to Figure

3.2, we note that though orbitals can contain electron pairs, all vacancies are occupied by single electrons before pairing occurs. Thus, p_x, p_y, and p_z orbitals will each be occupied by a single electron first. As the atomic number increases thereafter, pair formation will occur, as is seen in the transitions from carbon to oxygen shown in Figure 3.2.

Second, there is inherent stability to inert gas configurations. Thus, a shell with only a single electron in residence, a single electron in the s orbital, for example, tends to be energetically unstable and likes to contribute that electron elsewhere. On the other hand, a nearly complete shell, lacking one electron to completely fill the p orbitals, for example, will tend to be an eager acceptor of an electron. Referring again to Figure 3.2, we see that sodium contains a single electron in its 3s orbital while chlorine lacks one electron to complete its 3p orbitals, and we know that sodium readily contributes its electron (is oxidized) to chlorine (which is reduced) to form the ions and ionic bonds of sodium chloride. Note also that in the ionization process, sodium assumes the electron configuration of neon and chlorine the configuration of argon.

The inherent stability of the inert gas configuration gives rise to still another kind of interaction between atoms called covalent bonding. In a covalently bonded molecule, the valence electrons of each atom are redistributed within the molecule and shared among all the bonding atoms so as to complete orbital pairing and achieve an inert gas configuration about each.

Covalent bonding is illustrated in Figure 3.3 for molecular nitrogen. Note that sharing the 2p orbital electrons between the two nitrogen atoms completes pairing in each atom's p orbitals. It also completes the octet of an inert gas configuration about each atom, each sharing three electrons with the other to make a total of eight around each. Completion of the valence electron's octet in bonding is referred to as the Lewis Octet theory [Jolly 1984, 40] which states, in essence, that all chemical bonding strives to complete the inert gas octet, either by formal electron contribution as in ionic bonding or by sharing electrons as in covalent bonding.

A major distinction between covalent and ionic bonds, incidentally, resides in the mobility of the product entities involved. Ionic bonds are loose. The parties to the bond can behave essentially independently of one another. An ion of a pair can react, for example, with another entity, independently of its partner. Covalent bonds, on the other hand, are very strong. The product of a covalent bond is a new molecule with its own unique chemical charcacteristics.

In actual fact, however, no chemical bond is entirely ionic or covalent. Rather, all bonds exhibit both ionic and covalent character in varying degrees, ranging from something very close to all ionic to very close to

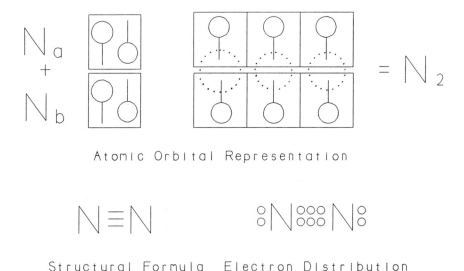

Atomic Orbital Representation

N≡N   oN ooo No
                                    o   ooo   o

Structural Formula   Electron Distribution

Figure 3.3   The nitrogen molecule, an example of covalent bonding. The p electrons of each of the atoms are shared between the two.

all covalent. Thus, even though covalency is bonding through shared rather than contributed electrons, one can still regard one of the bonded partners as an electron contributor and the other as an acceptor. In that sense, the acceptor may be thought of as reduced and the contributor as oxidized in the formation of the new entity.

Which of a bonded pair behaves as the electron contributor or acceptor depends upon relative factors like the nuclear charge and atomic size. These factors determine the element's electroncgativity, i.e., the strength with which it can attract electrons to itself. Values for the electronegativity of a number of elements are shown in Table 3.1 [Pauling 1948, 60]. Note that carbon is more electronegative than hydrogen, and oxygen is more electronegative than carbon. The values imply that hydrogen, in bonding with carbon, will behave as an electron donor, i.e., behave as if oxidized while carbon will behave as if reduced. On the other hand, carbon will behave as if oxidized in bonding with oxygen. Thus, carbon is regarded as reduced in methane and oxidized in carbon dioxide, despite the fact that bonding is essentially all covalent in both compounds.

The familiar structural formulas of organic chemistry can frequently be described in terms of the atomic orbital diagrams of Figures 3.2 and

Table 3.1 Electronegativity Values for a
Few Elements

Element	Electronegativity
Hydrogen	2.1
Carbon	2.5
Nitrogen	3.0
Oxygen	3.5
Fluorine	4.0
Silicon	1.8
Phosphorous	2.1
Sulfur	2.5
Chlorine	3.0
Germanium	1.8
Arsenic	2.0
Selenium	2.4
Bromine	2.8
Iodine	2.5

3.3.* An example is shown in Figure 3.4, where the structural formula
(b) and an electron configuration diagram of oxalic acid (c) are shown
along with an atomic orbital analogy (a). The single bond between the
carbons in the structural formula is shown as the sharing of two unpaired
carbon s electrons while the single and double bonds between carbon and
oxygen are shown as shared, unpaired p electrons.[†] Two of the oxygens
of the oxalate moiety then still remain with incomplete p orbitals that are
available for bonding; and these accept the s electrons of two atoms of
hydrogen, which complete the p orbitals and form the oxalate and hydrogen ions of oxalic acid. Note that sharing electrons in this fashion

---

*The use of atomic orbital arguments throughout this discussion is for illustrative purposes
only. They are not precise quantum mechanical descriptions of electron configurations about
molecules. Electron configurations about molecules are more rigorously described by molecular orbital theory, a topic beyond the scope of this book.
[†]The unpaired electrons in the 1s orbitals of carbon are a kind of contradiction to the rule
that orbitals pair electrons before higher levels are filled. In bonding, however, electrons
will frequently redistribute themselves in such a way as to be equi-energetic (see below in
the section on carbon chemistry).

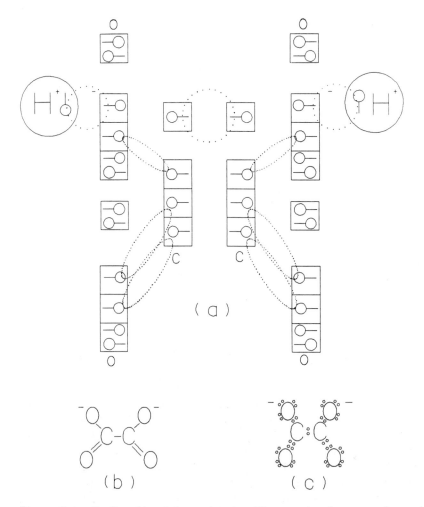

Figure 3.4   Oxalic acid and the oxalate ion. The negative charges on the oxalate result from the contribution of electrons by hydrogen atoms which, as a result, become hydrogen ions.

allows all the atoms of the oxalate ion to have an octet of electrons surrounding them.

In covalent bonds, incidentally, atoms may vibrate along and rotate about their bond axes. The frequencies of the vibrations and rotations depend upon the molecular groups engaged in the bonds and on the bond type. They are in the infrared region of the spectrum and serve as the basis of infrared absorption spectroscopy, an analytical method that is applicable to environmental investigation and will be discussed in more detail later.

CARBON CHEMISTRY

Elemental Carbon

The lowest energy electron configuration of elemental carbon's valence electrons is shown in Figure 3.2, where there is a pair shown in the 2s orbital and two unpaired electrons in the 2p orbitals. In bonding, however, carbon's electrons are distributed in a slightly different configuration. One of the 2s electrons is promoted to the vacant 2p orbital, and all the valence electrons become energetically equal. It is by virtue of these energy equivilent electrons that carbon can assume the apparent oxidation states between -4 and $+4$ that are shown in Table 3.2.

Despite the apparent multiple oxidation states, however, carbon nearly always behaves as a neutral atom. Its four outer electrons lie midway between the inert gas configurations of helium and neon, and energy considerations dictate extreme difficulty for it to either accept four electrons from another element to form a C^{4-} ion or contribute them to another element to form a C^{4+} ion. The difficulty arises because as electrons are removed from or added to an atom's outer orbitals, the overall electronic charge on the resultant species increases, making the removal or addition of more electrons progressively harder. Only very large atoms like tin and lead, which also have four electrons in their outer orbitals, can form M^{4+} ions, and they do so by virtue of their size. That is, the very large electronic clouds around their nuclei tend to shield their

Table 3.2   Carbon Oxidation States

Oxidation state	Formula	Chemical name
C^{4-}	CH_4	Methane
C^-	CaC_2	Calcium carbide
C^0	C	Graphite, diamond
C^{2+}	CO	Carbon monoxide
C^{3+}	$C_2O_4^{2-}$	Oxalate ion
C^{4+}	CO_2	Carbon dioxide
C^{4+}	HCO_3^-	Bicarbonate ion
C^{4+}	CO_3^{2-}	Carbonate ion
C^{4+}	CS_2	Carbon disulfide

nuclear charges, making removal of electrons easier than with small atoms like carbon. Carbon always, therefore, forms covalent bonds, sharing its four outer electrons with other elements and with itself. Even in materials in which carbon exhibits ionic character, it does so by formation of a complex of carbon atoms covalently bonded to one another like in calcium carbide, the structure of which is shown in Figure 3.5. It is carbon's remarkable ability to form stable covalent bonds with itself that is the basis of the formation of the countless numbers of biological and organic carbon compounds that exist. And its apparent oxidation state in all chemical compounds derives from its ability to serve as either an electron

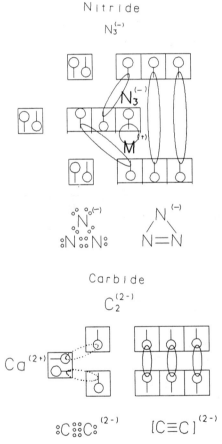

Figure 3.5   The carbide and nitride ions. Note that they are complex ions composed of covalently bonded atoms.

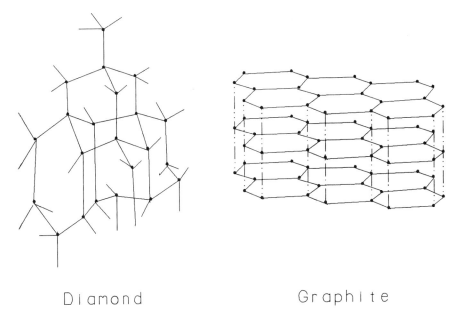

Diamond   Graphite

Figure 3.6   Diamond and graphite structures.

contributor or acceptor in terms of its electronegativity relative to its
bonding partners.

Elemental carbon exists in two crystalline forms, graphite and diamond, the structures of which are illustrated in Figure 3.6. The diamond
structure is that of a covalently bonded giant molecule of carbon atoms,
with each carbon sharing its electrons in single bonds with four others.
The structure of graphite is that of layers of six-membered rings, in which
each carbon is covalently bound to three others, with the fourth electron
in each serving to bind the layers together. Charcoals, coke, coals, and
lampblacks are impure forms of graphite.

When heated to high temperatures in the absence of oxygen, carbon
can unite with many elements including calcium, hydrogen, and sulfur to
form calcium carbide, methane, and carbon disulfide, respectively. Calcium carbide reacts with water to form acetylene

$$CaC_2 + 2H_2O = Ca(OH)_2 + CH{\equiv}CH$$

and carbon disulfide burns in air to form sulfur dioxide

$$CS_2 + 3O_2 = CO_2 + 2SO_2$$

Carbon disulfide may also be used to prepare carbon tetrachloride by reacting it with chlorine gas

$$CS_2 + 3Cl_2 = CCl_4 + S_2Cl_2$$

Carbon Dioxide and Carbonates

When heated in an excess of oxygen, carbon burns and forms carbon dioxide. When carbon dioxide is dissolved in water, the weak divalent carbonic acid is formed

$$CO_2 + H_2O = H_2CO_3$$

In nearly neutral (pH = 6.4) solutions, carbonic acid dissociates and forms bicarbonate ions while in basic solutions (pH = 10), it dissociates to carbonate ions. Thus,

$$\text{pH increasing} \rightarrow$$
$$H_2CO_3 = H^+ + HCO_3^- = H^+ + CO_3^{2-}$$

But note, in keeping with our discussion above, that carbonate and bicarbonate ions are not carbon ions. Rather, they are ions of covalently bonded carbon-oxygen entities, the ionic charge resulting from the electrons left in the oxygens' p orbitals by hydrogen as it dissociates to form hydrogen ions (Figure 3.7).

Potassium or sodium salts in solution with carbonic acid form bicarbonates according to

$$Na^+ + HCO_3^- = NaHCO_3$$

and heating bicarbonates releases carbon dioxide and forms the carbonates

$$2NaHCO_3 + \text{heat} = Na_2CO_3 + CO_2 + H_2O$$

Heating calcium and magnesium carbonates releases carbon dioxide and forms the oxides, i.e.,

$$CaCO_3 + \text{heat} = CaO + CO_2$$

and acid treatment of carbonates and bicarbonates also frees carbon dioxide. Thus,

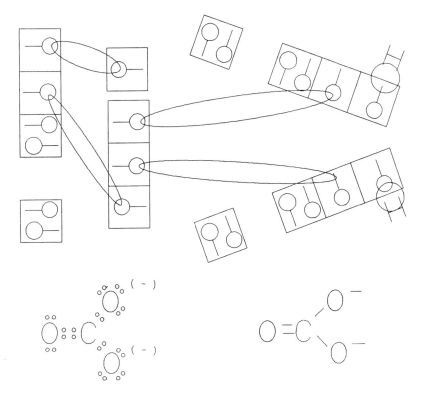

Figure 3.7   The carbonate ion. The negative charges on the oxygens derive from the contribution of electrons from positively charged counter ions like hydrogen.

$$HCl + NaHCO_3 = CO_2 + NaCl + H_2O$$

and

$$2HCl + CaCO_3 = CO_2 + CaCl_2 + H_2O$$

Carbon Monoxide

If carbon is burned in a limited supply of oxygen, the reaction product is carbon monoxide rather than carbon dioxide. Passing steam over hot coke will also produce carbon monoxide but in a mixture that includes hydrogen

$$H_2O(\text{steam}) + C(\text{coke}) = H_2 + CO$$

The mixed gases thus produced are sometimes called "synthesis gas" because they may be used with appropriate catalysts to produce complex organic compounds.

Carbon monoxide itself is combustible and is oxidized to carbon dioxide if burned in oxygen. It may also be catalytically reduced back to carbon at elevated temperatures.

Carbon monoxide binds strongly to transitional metals (those atoms filling their d orbitals like iron and cobalt) to form a class of compounds called metal carbonyls. Its ability to do so is the reason it is poisonous, since it binds preferentially to the iron atom in hemoglobin and thereby inhibits oxygen transport in the circulatory system.

Organic Carbon

Some common organic carbon compound structures are illustrated in Figure 3.8. Straight chains of carbon and hydrogen are classified as paraffins while ring compounds like benzene are classified as aromatics. Organic derivatives of the basic paraffins and aromatics are those with functional groups attached to them to make alcohols, aldehydes, carboxylic acids, or sulfur and nitrogen derivatives like amines and mercaptans.

The most reduced forms of carbon are those found in organic compounds where carbon is mostly bonded to hydrogen, although it is also commonly bonded to other elements as well. Organic carbon compounds almost universally burn in an oxygen environment and the reaction products are carbon dioxide, the most oxidized form of carbon, and water. And like elemental carbon, if the oxygen supply is limited, the reaction products are carbon monoxide and water. As will be seen later, the combustion of carbon compounds, as well as acid treatment of carbonates, to produce carbon dioxide, serves as the basis of many analytical methods used to monitor carbon pollutants in air and water.

In general, organic carbon compounds range between insoluble and only very slightly soluble in water. However, many compounds that have been partially oxidized like organic acids, alcohols, some aldehydes, some amines, and others with ionic character like sulfonic acids do exhibit water solubility. In the process of biological decay, biomolecules and organic compounds can pass through oxidation sequences that carry them from reduced carbon compounds with limited water solubility to soluble oxidized forms, which accounts for the presence of soluble organic matter in polluted waters.

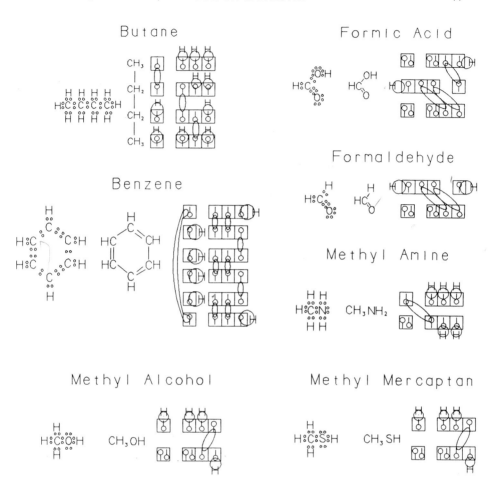

Figure 3.8   Some examples of organic carbon structures and their atomic orbital representations.

NITROGEN CHEMISTRY

Elemental Nitrogen

In contrast to carbon's ability to form complex, many-atom molecules by covalently bonding to itself, nitrogen chains rarely exceed two. That nitrogen exhibits such behavior is attributed to the pair of unbonded electrons shown residing in the 2s orbital in Figure 3.2. These two electrons, which are frequently referred to as lone pairs, serve to mutually repel nitrogen atoms from one another, causing single bonds, in particular, to be weak, and making long nitrogen chains unstable. Thus, while elemental carbon occurs in the forms of the giant molecules of diamond and graphite, elemental nitrogen occurs only as a bi-atomic gas molecule at normal temperatures and pressures.

The remarkable differences in the physical and chemical properties of the hydrides of nitrogen and carbon, ammonia and methane, respectively, are also attributable to the lone pair electrons. Although methane is a relatively unreactive neutral compound that liquefies at very low temperature ($-164°C$), ammonia is a strong base, is relatively reactive, and liquefies at $-33°C$. Thus, the lone pair serves to enhance molecular association, to increase reactivity, and as a potential partner with a hydrogen ion in an acid-base interaction.

Nitrogen's apparent oxidation states range from 3^- in ammonia to 5^+ in nitrates and are a consequence of the five electrons in its 2s and 2p orbitals. Similar to carbon, its bonding is primarily covalent, and it engages in covalent bonds with carbon in countless organic compounds. It also forms covalently bonded complex ions. It can, for example, combine with alkali metals like sodium to form azides, ionic materials containing the nitride ion (N_3^-). It is interesting to note that the apparent oxidation state of nitrogen in the nitride ion, which is a complex of three covalently bonded nitrogen atoms, is $-1/3$ (Figure 3.5).

Ammonia

Ammonia, as we have discussed before, is a product of the biodegradation of proteins and amino acids. It represents the most reduced form of nitrogen. It is soluble in water to the extent of 30% by weight, but exhibits high volatility at high concentrations, accounting for the irritating odor of its concentrated solutions.

In water, a small amount of dissolved ammonia reacts to form ammonium hydroxide

$$NH_3 + H_2O = NH_4^+ + OH^-$$

The remainder forms an ammonia hydrate of the form $NH_3 \cdot H_2O$ [Jolly 1964, 40].

Ammonia is the starting material for the synthesis of nitrogenous fertilizers. It is manufactured, using the Haber process, by reacting atmospheric nitrogen with hydrogen at pressures of about 1000 atmospheres and at about 500°C in the presence of a catalyst.*

Ammonia readily forms very water soluble ammonium salts by reaction with acids. Thus,

$$NH_3 + HCl = NH_4Cl$$

or

$$NH_4OH + HCl = NH_4Cl + H_2O$$

and

$$NH_4OH + HNO_3(\text{dilute and cold}) = NH_4NO_3 + H_2O$$

Heating ammonium salts with oxidizing anions decomposes them to nitrogen or nitrogen oxides according to

$$NH_4NO_2 = N_2 + H_2O$$

and

$$NH_4NO_3 = N_2O + H_2O$$

and aqueous ammonium ions are completely oxidized to nitrogen and nitrogen oxides by hot mixtures of nitric and hydrochloric acids.

Oxides of Nitrogen

There are a wide variety of oxides of nitrogen. Most important from an environmental viewpoint are nitrous oxide (N_2O), nitric oxide (NO), nitrogen dioxide (NO_2), and nitric acid (HNO_3) in which the apparent oxidation states of nitrogen are +1, +2, +4, and +5, respectively. These are the oxides that are intimately involved in smog formation, acid rain, stratospheric ozone depletion, ground water contamination, and water eutrophication.

---

*The hydrogen used in the process is generally produced by passing steam over hot coke to form synthesis gas from which the carbon monoxide is removed.

Nitrous Oxide

Nitrous oxide is second to nitrogen in atmospheric abundancy, being generated by denitrifying bacteria, as has been discussed earlier. It may also be formed in the atmosphere by reaction with ozone

$$N_2 + O_3 = N_2O + O_2$$

Of all the oxides of nitrogen, N_2O is the least reactive with respect to oxidation-reduction propensity. It is, however, subject to thermal decomposition at lower temperatures than the other nitrogen oxides. Thus, heated to about 500°C, it decomposes rapidly producing nitric oxide

$$4N_2O = 3N_2 + O_2 + 2NO$$

A similar decomposition reaction also occurs in the upper atmosphere where N_2O may be photodecomposed by ultraviolet radiation from the sun, and where the nitric oxide produced enters into reaction with the ozone layer (see below).

Because of its low thermal decompositon temperature, nitrous oxide supports combustion. A glowing splint immersed in it will burst into flame because the splint temperature is high enough to decompose the N_2O, providing sufficient oxygen to make the burning reaction self perpetuating.

The simplest atomic orbital representation of nitrous oxide is illustrated in Figure 3.9a. Structural studies of the molecule, however, indicate that it is not the ring structure shown. Rather, it appears to be a linear molecule with its bonding electrons resonating* between the two structures shown in Figure 3.9b [Moeller, 1952, 592]. In the structure on the left in Figure 3.9b, the atomic orbital representation shows the unpaired electrons of the central nitrogen atom as pairing with the unpaired electrons of the second nitrogen atom, forming the triple bond. The p electrons of the oxygen atom are shown as paired, thus leaving one p orbital vacant. The bond between the oxygen and the central nitrogen is then made by nitrogen's contribution of its *paired* 2s electrons to the vacancy.

It must be mentioned at this point that the bond formed as a conse-

---

*The bonding electrons of all molecules resonate between configurational extremes like these. The atomic orbital representations used in this discussion are, thus, simply examples of selected possibilities.

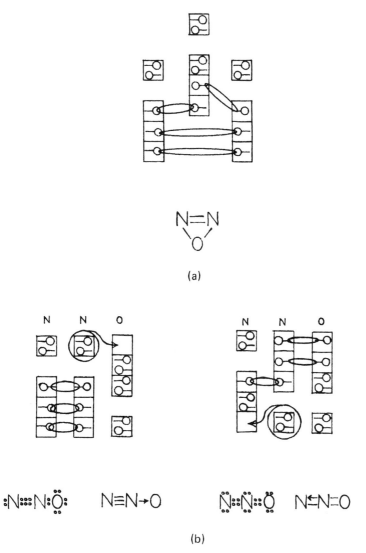

Figure 3.9 (a) A simplified but incorrect atomic orbital representation of nitrous oxide; (b) Resonance structures of nitrous oxide that represent its molecular structure as determined experimentally (see text). Note the "coordinate covalent" bonds in the atomic orbital diagrams.

quence of the contribution of paired electrons by one atom to another is not the same as a single bond, despite the fact that it involves only two electrons. It is, therefore, sometimes called a "coordinate covalent" bond to distinguish it from a single bond. Single bonds are always combinations of *unpaired* electons from each of the bonding partners, and exhibit different bond energies and bond lengths than coordinate covalent bonds do. So a pair-contribution bond is sometimes denoted in structural formulas as an arrow [Sidgwick 1950, I 682]. In our examples, we have used arrows.

The atomic orbital representation of the second structure of nitrous oxide shows a coordinate covalent bond being made by contribution of the central nitrogen's pair of 2s electrons to the second nitrogen instead of the oxygen, and a single bond being made between two unpaired p electrons of each nitrogen. And finally, a double bond is shown between the remaining two of the central nitrogen's unpaired p electrons and the unpaired p electrons of the oxygen. The structural formula shown for this resonant form of nitrous oxide is after Sidgwick [1950, I 682].

Nitric Oxide

Nitric oxide, in contrast to nitrous oxide, can be formed from the elements at elevated temperatures. Thus, burning coal-gas or hydrogen in air produces NO from atmospheric oxygen and nitrogen, as does burning hydrocarbons in internal combustion engines. It may also be prepared by oxidation of ammonia, but requires a platinum catalyst to do so.

Nitric oxide is the most thermally stable of all the oxides of nitrogen, decomposing appreciably only at temperatures in excess of 1000°C. However, it can serve as an oxidizing agent at elevated temperatures and support the combustion of materials such as phosphorous. It may also, itself, be oxidized to nitrate by strong oxidizing agents like potassium permanganate.

Nitric oxide is a colorless gas that readily combines with oxygen at room temperature to form the redish brown nitrogen dioxide. In fact, it is often mistakenly thought that nitric acid dissolution of metals conducted in open vessels produces nitrogen dioxide. In reality, it is nitric oxide that is produced, which then reacts with atmospheric oxygen to form the dioxide [Jolley 1964, 74]. Thus

$$3Cu + 8HNO_3 = 3Cu(NO_3)_2 + 4H_2O + 2NO$$

and thence

$$2NO + O_2 = 2NO_2$$

Nitric oxide also reacts with ozone to form nitrogen dioxide, one of the reactions involved in ozone depletion in the upper atmosphere. The reaction is chemiluminescent, i.e., the nitrogen dioxide is initially produced in an excited electron state that emits light as it relaxes. The reaction is written

$$NO + O_3 = NO_2^* + O_2 = NO_2 + O_2 + photon$$

where the asterisk represents the excited nitrogen dioxide molecule.

The chemiluminescence associated with the reaction of ozone and nitric oxide, incidentally, serves as the basis of nitric oxide detection and measurement in a number of commercial analytical devices to be discussed in more detail later.

While nitrous oxide is water soluble to the extent of about 130 cc per 100 cc of water, nitric oxide is only sparingly soluble (about 7.5 cc per 100 cc H_2O). It also has a much lower boiling point than nitrous oxide, $-151°C$ compared to $-88°C$, implying a lower tendency for molecular association than that of nitrous oxide.

The atomic orbital electron configuration of nitric oxide is shown in Figure 3.10. Note that the molecule has an odd number of electrons, five

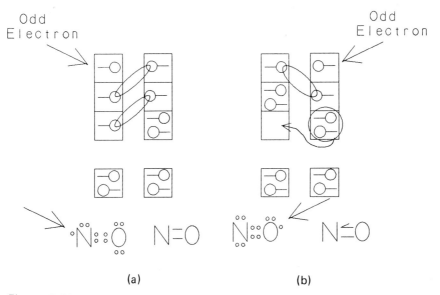

Figure 3.10 Nitric oxide showing the resonance configurations that comprise a "three-electron bond."

from nitrogen and six from oxygen. After covalently bonding to oxygen, the nitrogen atom remains with a single unpaired electron. Nitric oxide is, as such, classified as an "odd molecule" [Pauling 1948, 266]. According to Pauling, when two atoms of an odd molecule are similar in size and electronegativity, as are oxygen and nitrogen, neither the filled orbital of the oxygen or the unpaired electron of the nitrogen can be thought of as being localized on either atom. Rather, the electrons resonate between the two possible structures shown in Figure 3.10a and 3.10b and, thereby, form a "three-electron bond" with a bond strength about half that of a normal single bond.

Nitrogen Dioxide

Nitrogen dioxide is similar to nitric oxide in that it is also an odd molecule, and it has a three-electron bond in its structure (compare Figure 3.11a with Figure 3.10). Nitrogen dioxide, however, can couple to itself. At normal temperatures and pressures, it is always in equilibrium with its dimer, nitrogen tetroxide,

$$2NO_2 = N_2O_4$$

The dimer forms because, relative to nitric oxide, the three-electron bond of nitrogen dioxide is weak. From the viewpoint of atomic orbitals, the structure with the unpaired electron on the nitrogen atom (right side of Figure 3.11a) is favored. That unpaired electron bonds to a second nitrogen dioxide molecule to form the dimer (Figure 3.11b).

Like nitrous oxide, nitrogen dioxide may be thermally decomposed

$$2NO_2 = 2NO + O_2$$

but it is more stable than nitrous oxide. Appreciable decomposition starts only above 600°C.

Nitrogen dioxide reacts with water leading to the formation of both nitric and nitrous acids. The nitrous acid, however, is unstable and decomposes to form nitric oxide. The net reaction, which is

$$3NO_2 + H_2O = 2HNO_3 + NO$$

serves as the basis of the commercial production of nitric acid, the nitrogen dioxide being derived from the oxidation of ammonia (see above).

Nitrogen dioxide may also be photodecomposed by ultraviolet radiation

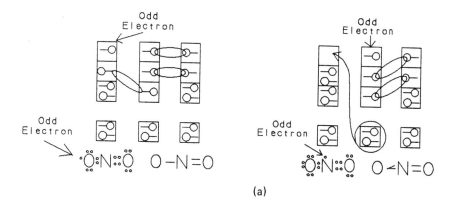

(a)

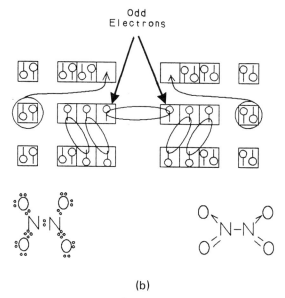

(b)

Figure 3.11 (a) The three-electron bond resonance structure of nitrogen dioxide; (b) The dimer of nitrogen dioxide (see text).

$$NO_2 + light = NO + atomic\ oxygen$$

with, as will be seen later, the atomic oxygen thus produced being actively associated with smog formation.

Nitrates and Nitric Acid

The apparent oxidation state of nitrogen in nitric acid and in nitrates is +5. The nitrates may be obtained by reacting metals or metal oxides with the acid. All normal nitrates are water soluble as is nitric acid itself, which is miscible with water in all proportions.

Although sodium and potassium nitrates are thermally stable, they will decompose to the nitrite salts and oxygen if they are heated to red heat. Thus

$$2NaNO_3 = 2NaNO_2 + O_2$$

Other metal nitrate salts decompose to the oxides of the metals, nitrogen dioxide, and oxygen when they are heated. Thus,

$$2Cu(NO_3)_2 = 2CuO + NO_2 + O_2$$

Ammonium nitrate, on the other hand, produces nitrous oxide and water when heated gently, but can decompose explosively if heated vigorously.

Alkaline solutions of nitrates may be reduced to ammonia by aluminum and zinc, a property sometimes used for the quantitative analysis of nitrates.

Nitric acid is a strong oxidizing agent, not only converting metals to their salts, but acting on nonmetals like carbon and sulfur to form their oxides. Thus,

$$C + 4HNO_3 = 4NO_2 + 2H_2O + CO_2$$

and

$$S + 6HNO_3 = 6NO_2 + 2H_2O + SO_2$$

It also reacts with alcohols to form nitrates and with aromatic organics like benzene to yield nitro-compounds with reactions of the form

$$C_2H_5OH + HNO_3 = C_2H_5(NO_3) + H_2O$$

and

$$C_6H_6 + HNO_3 = C_6H_5(NO_2) + H_2O$$

And interaction between dilute nitric acid and sulfides yields free sulfur

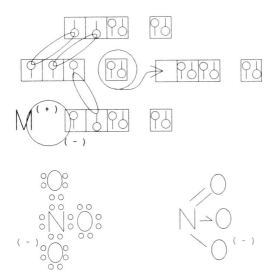

Figure 3.12   The nitrate ion.

and nitric oxide, so

$$3H_2S + 2HNO_3 = 3S + 2NO + 4H_2O$$

Upon exposure to light, nitric acid slowly decomposes to nitrogen dioxide, oxygen, and water. The tendency for old nitric acid solutions to be slightly yellow results from the solution of NO_2 in them.

One possible resonance structure for the nitrate ion is shown in Figure 3.12. Consideration of atomic orbitals suggests the following redistribution of electrons to achieve the bonding arrangement. First, the electron contributed by the formation of the positive ion is contributed to one of the oxygens, imparting a negative charge to it. Second, nitrogen contributes two of its unpaired p electrons to a double bond with one of the uncharged oxygens and its remaining unpaired p electron to the charged oxygen, forming a single bond with it. Finally, nitrogen contributes its paired 2s electrons to the third oxygen, forming a coordinate covalent bond with it.

SULFUR CHEMISTRY

Elemental Sulfur

Elemental sulfur's electron configuration is shown in Figure 3.2. As can be seen, its first two electron shells are full, and its valence electrons reside in its 3s and 3p orbitals, having the same configuration as oxygen's. As a consequence, its behavior mimics oxygen within limits determined by its atomic radius, which is 40% larger than oxygen's. Like oxygen, therefore, it forms ionic bonds with an apparent oxidation state of 2^-. But in contrast to oxygen, it can exhibit multiple oxidation states, ranging from 2^- in sulfides to 6^+ in sulfuric acid, and it can bond to itself in stable cyclic and chain molecules of large size. Oxygen is limited to bi- or tri-atomic molecules, like O_2 and O_3, and to a single oxidation state.

Oxygen's limitations derive from its small size and high electronegativity. Its tight electron cloud, like the lone pair electrons of nitrogen, produces high interatomic repulsions, inhibiting extended oxygen structures, and its high electronegativity makes it a strong electron acceptor but not a donor. Sulfur, on the other hand, being less electronegative and larger, is able to contribute as well as accept electrons and engage in multiple bonds with itself.

Sulfur is a solid at room temperature and boils at about 444°C. It has two crystalline forms, rhombic and monoclinic. Rhombic sulfur is its most stable form at normal temperatures, converting to monoclinic sulfur above 96°C. Rhombic sulfur melts at 112°C. The monoclinic form melts at 119°C. But on solidifying from the liquid, both forms usually crystallize, so the observed melting temperature frequently varies between the two extremes, depending upon the proportions of each present.

Below 800°C, sulfur normally occurs, regardless of whether it is in the solid, liquid, or vapor states, as an eight atom molecule configured as a puckered octagon (Figure 3.13) [Sidgwick 1950, II 876]. Above 800°C, it is mostly the di-atom structure, S_2, and above about 2000°C, it is mostly in the form of single atoms.

If liquid sulfur is heated, its viscosity increases to a maximum at 159°C. This property is attributed to opening the eight unit ocatagonal molecules

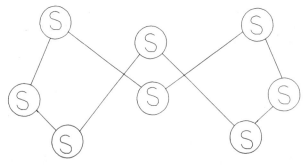

Figure 3.13   The structure of the eight membered sulfur molecule.

to form long chains of polymeric sulfur constucted of as many as 10^5 eight unit linear molecules [Meyer 1968, 248]. Above the maximum, the polymer begins to revert to the eight membered octagon. A plastic, amorphous form of sulfur, which may be obtained by rapidly cooling the liquid from the temperature of the viscosity maximum, is thought to be a supercooled liquid form of the polymeric sulfur.

Sulfur is a relatively reactive element resembling oxygen in its behavior. Thus, those substances that readily combine with oxygen usually combine with sulfur as well, e.g., forming sulfides analogous to oxides

$$2Cu + S = Cu_2S$$

and

$$Fe + S = FeS$$

and

$$2C + S = CS_2$$

and forming hydrides analogous to water

$$S + H_2 = H_2S$$

Burning sulfur in oxygen, of course, forms sulfur dioxide, a situation in which the sulfur contributes its electrons to the more electronegative oxygen

$$O_2 + S = SO_2$$

And sulfur may be oxidized to sulfuric acid by strong oxidizing agents like nitric acid

$$S + 6HNO_3 = H_2SO_4 + 6NO_2 + 2H_2O$$

Sulfur Hydrides and Sulfides

The hydrides of sulfur include hydrogen sulfide and the class of organic compounds called mercaptans. The mercaptans are the sulfur analogs of the alcohols, e.g.,

$$Methyl\ Alcohol = CH_3OH$$

and

Methyl Mercaptan = CH_3SH

The sulfur-hydrogen group in mercaptans, as well as in related compounds like the amino acid cysteine discussed in Chapter 1, is referred to as a sulfhydryl group.

Hydrogen sulfide occurs naturally. It is found commonly in volcanic emissions as well as in solution in mineral waters and hot springs associated with volcanic activity, and it is frequently associated with petroleum and natural gas deposits. It is also a product of the anaerobic decompostion of animal matter and sulfates. It is very poisonous, approaching the toxicity of cyanide gas.

The gas liquifies at $-60.4°C$ and liquid H_2S solidifies at $-85.5°C$. Thus, despite the fact that the electron configurations of water and hydrogen sulfide are similar, the physical characteristics of each are quite different. The decreased molecular association implied by the lower melting and boiling points of hydrogen sulfide, compared to water, is attributed to a lower degree of hydrogen bonding in the former compared to the latter.* Water is a strongly hydrogen bonded substance because of the high electronegativity of its oxygens. Thus, it is a highly associated substance that exhibits high melting and boiling points. Because of the size of the sulfur atom, however, it is less electronegative than oxygen and forms weaker hydrogen bonds. Thus, it is a less associated substance than water with a lower melting point and a lower boiling point.

A common synthetic source of hydrogen sulfide is through treatment of inorganic sulfides with dilute hydrochloric acid

$$FeS + 2HCl = H_2S + FeCl_2$$

The gas can serve as a reducing agent, reacting with sulfuric acid to form sulfur dioxide and elemental sulfur

$$H_2S + H_2SO_4 = SO_2 + S + 2H_2O$$

and it burns in air to form sulfur dioxide and water

$$2H_2S + 3O_2 = 2SO_2 + 2H_2O$$

---

*Hydrogen bonding is a physical attraction between the less electronegative hydrogens of one molecule and the more electronegative atoms of another. It frequently leads to the physical association of the two molecules.

It is soluble in water, forming a weak divalent acid that ionizes according to

$$\text{pH increasing} \rightarrow$$
$$H_2S = H^+ + HS^- = H^+ + S^{2-}$$

and which may be neutralized by metal hydroxides to form both hydrosulfides and sulfides

$$H_2S + NaOH = NaHS + H_2O$$

and

$$NaHS + NaOH = Na_2S + H_2O$$

If elemental sulfur is dissolved in a solution of soluble sulfide or hydrosulfide, polysulfides and polyhydrosulfides of varying composition, e.g., from Na_2S_2 to Na_2S_5, may be formed, the bonding in which is thought to be through covalent attachments of sulfurs to one another. Polysulfides may also be formed from aging or oxygenating sulfide and hydrosulfide solutions [Chen 1974, 109] and are, therefore, sometimes components of impounded waters. Acid treatment of polysulfides releases elemental sulfur and hydrogen sulfide.

Oxides of Sulfur

In addition to the commonly occuring sulfur dioxide, a number of other sulfur oxides are known to exist ranging from disulfur monoxide (S_2O) and sulfur monoxide to polysulfur oxides of the form S_nO_2. With the exceptions sulfur dioxide and sulfur trioxide, however, most are either unstable or academic curiosities and will not be discussed here.

Sulfur Dioxide

Sulfur dioxide is a colorless gas with an irritating and pungent odor that may be produced in a number of ways including combustion of sulfur in air*

$$S + O_2 = SO_2$$

by roasting metallic sulfides in air

---

*A small amount of sulfur trioxide is also produced when sulfur is burned in air or oxygen.

$$4FeS_2 + 11O_2 = 2Fe_2O_3 + 8SO_2$$

by treating a soluble sulfite with acid

$$2HCl + Na_2SO_3 = SO_2 + 2NaCl + H_2O$$

or by the reduction of sulfuric acid with metals, carbon, or sulfur

$$Cu + 2H_2SO_4 = CuSO_4 + 2H_2O + SO_2$$
$$C + 2H_2SO_4 = CO_2 + 2H_2O + 2SO_2$$
$$S + 2H_2SO_4 = 2H_2O + 3SO_2$$

The gas is heavier than air, liquefies at relatively high temperature ($-10°C$), and freezes at $-72°C$.

Sulfur dioxide is a stable compound, but it can be decomposed by heating above 1000°C or by irradiating with ultraviolet light. The general reaction is thought to be of the form

$$4SO_2 + energy = 4SO_2^* = 4SO + 4O = 2S + 2SO_3 + O_2$$

where the asterisk indicates an excited sulfur dioxide molecule that breaks into sulfur monoxide and oxygen atoms, which themselves react to form elemental sulfur, sulfur trioxide, and oxygen [Schenk and Steudel 1968].

Sulfur dioxide behaves primarily as a reducing agent, being easily oxidized to sulfuric acid by reaction with oxygen in water

$$2SO_2 + O_2 + 2H_2O = 2H_2SO_4$$

but it can also behave as an oxidizing agent, oxidizing metal ions in acid solutions, itself being reduced to elemental sulfur

$$4M^+ + SO_2 + 4H^+ = 4M^{2+} + 2H_2O + S$$

Studies of the sulfur dioxide molecule suggest that its electronic structure is that shown in Figure 3.14. In the atomic orbital representation, the unpaired electrons of the sulfur atom are shown as pairing with the unpaired electrons of one of the oxygens to form the double bond shown in the electronic structure. All the electrons of the second oxygen are shown as paired, the bond between it and sulfur being made by sulfur's contribution of two *paired* electrons to that oxygen's vacant 2p orbital to form a coordinate covalent bond with it.

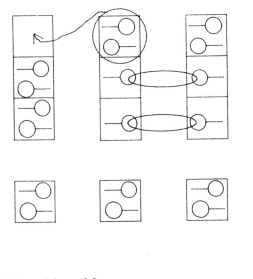

Figure 3.14 Sulfur dioxide.

Sulfur dioxide is quite water soluble. At standard temperature and pressure, a saturated solution contains about 53% SO_2 by weight [Schenk and Steudel 1968]. In solution, a small amount of the SO_2 reacts with water to form sulfurous acid, a weak diprotic acid that can be neutralized with alkali hydroxides to form bisulfites and sulfites, but does not, itself, exist out of solution. Thus,

$$SO_2 + H_2O = H_2SO_3$$

and

$$H_2SO_3 + NaOH = NaHSO_3 + H_2O$$

and

$$NaHSO_3 + NaOH = Na_2SO_3 + H_2O$$

Sulfurous acid can behave as both an oxidizing and reducing agent. It is easily oxidized by atmospheric oxygen to form sulfuric acid

$$2H_2SO_3 + O_2 = 2H_2SO_4$$

and it will oxidize hydrogen sulfide to elemental sulfur

$$2H_2S + H_2SO_3 = 3H_2O + 3S$$

In contrast to sulfurous acid, sulfite salts can be isolated in a free state. They are, however, decomposed by heat, regenerating sulfur dioxide and forming sulfates and sulfides

$$2NaHSO_3 = SO_2 + Na_2SO_3 + H_2O$$

and

$$4Na_2SO_3 = Na_2S + 3Na_2SO_4$$

And, like their parent acid, they are easily oxidized, forming their corresponding sulfates.

By boiling sodium sulfite with sulfur, one obtains sodium thiosulfate

$$Na_2SO_3 + S = Na_2S_2O_3$$

an important analytical reagent that can serve as a primary standard for the quantitative determination of iodine, a common analytical procedure frequently applied in environmental monitoring. Reaction of the thiosulfate with intensely colored iodine produces colorless sodium iodide and sodium tetrathionate

$$2Na_2S_2O_3 + I_2 = 2NaI + Na_2S_4O_6$$

and, thus, provides a visual means to determine reaction completion.

Sulfur Trioxide

Sulfur dioxide may be catalytically oxidized to sulfur trioxide. Thus, in the presence of vanadium pentoxide

$$2SO_2 + O_2 = 2SO_3$$

Sulfur trioxide is a powerful dehydrating agent. It is also an oxidizing agent, converting elemental sulfur to sulfur dioxide at about 100°C and oxidizing moist hydrogen sulfide to elemental sulfur. At elevated temperatures or under the influence of ultraviolet radiation, it decomposes to sulfur dioxide and oxygen.

The catalytic oxidation of sulfur dioxide to sulfur trioxide serves as the basis of the manufacture of sulfuric acid. That is, the SO_3 produced is dissolved in water to form the acid.

$$SO_3 + H_2O = H_2SO_4$$

Sulfuric acid itself is also very water soluble, generating a great deal of

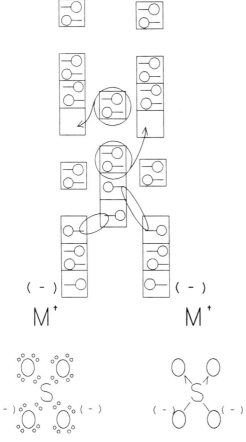

Figure 3.15   The sulfate ion. The bisulfate ion would have a hydrogen substituted for one of the "M^+" ions.

heat as it dissolves. And, like its anhydrous precursor, sulfur trioxide, it is also a powerful dehydrating agent, forming hydrates when dissolved in water that vary from a monohydrate ($H_2SO_4.H_2O$) at about 85% by weight of SO_3 to an octahydrate ($H_2SO_4.8H_2O$) at about 35% SO_3 [Barnett and Wilson 1959, 494].

Sulfuric acid is also a strong diprotic acid, ionizing in water to form both the bisulfate and sulfate ions (Figure 3.15)

$$H_2SO_4 = H^+ + HSO_4^- = H^+ + SO_4^{2-}$$

Its neutralization by alkali hydroxides produces the corresponding soluble ionic salts

$$H_2SO_4 + NaOH = NaHSO_4 + H_2O$$

and

$$NaHSO_4 + NaOH = Na_2SO_4 + H_2O$$

But sulfate ions also form insoluble salts with a number of metal ions, one of the most analytically important being barium. That is, barium sulfate, being an extremely water insoluble salt, serves as the basis of a commonly used gravimetric determination of sulfate.

THE BIOCHEMISTRY OF CARBON, NITROGEN, AND SULFUR

Implicit in the discussion of the movement of our subject elements among the regions of the biosphere are the roles they play in living organisms. In the following discussion, those roles are discussed in the context of some elementary biochemical principles, with emphasis on subjects like photosynthesis, nitrogen fixation, and respiration, since it is by these means that carbon, nitrogen, and sulfur pass back and forth between living organisms and the environment.

Cells and Their Contents

Living organisms are constructed of cells, i.e., small membrane bounded vesicles containing water, a large number of dissolved chemicals, and a number of suspended structures, all of which serve in one fashion or another to help maintain cell viability. But these cell constituents are actually an ordered hierarchy of biomolecules of increasing molecular complexity that are born from water and our subject elements in the form of carbon dioxide, molecular nitrogen, and sulfates, all obtained from the environment. Our subject elements then, in all their complex biomolecular forms, cycle within and among the planet's diversity of organisms and ultimately return again to the environment as simple small molecules, but now generated as end products of life processes.

Within the cell, the simple starting molecules are converted in a series of metabolic steps into intermediates of increasing molecular size, which ultimately link together to form macromolecules like proteins, nucleic acids, polysaccharides, and lipids, structural chemical formulas for which are illustrated in Figure 3.16. The macromolecules, at the next level of organization, combine to form supra-molecular complexes like lipoproteins, which are loosely bonded complexes of lipids and proteins, or ribosomes, which are complexes of nucleic acids and proteins. At the highest level of organization, the supracomplexes assemble into the suspended matter of the cell, the organelles like the nuclei and mitochondria, or the chloroplasts of plant matter.

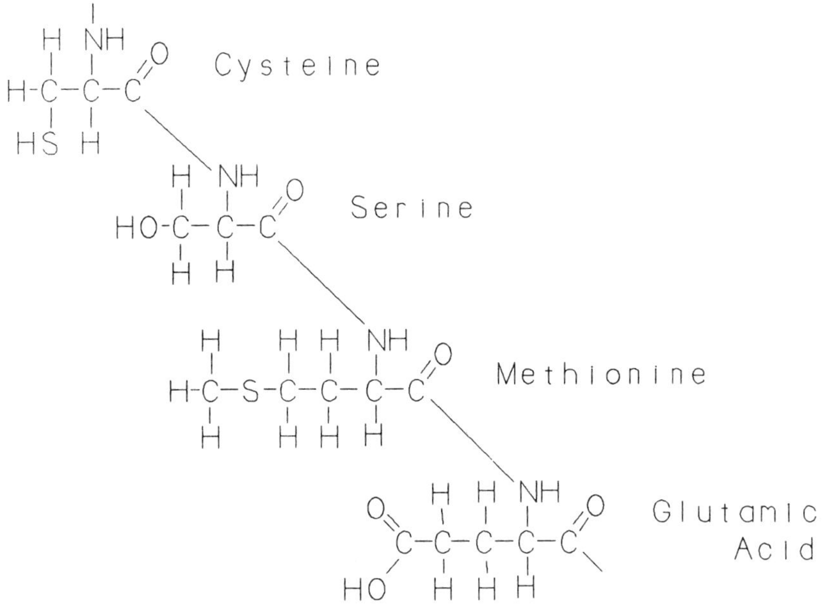

(a)

Figure 3.16 (a) A protein fragment showing four amino acids linked by peptide bonds; (b) four nucleotide bases linked through phosphate groups to form a short segment of deoxyribonucleic acid (DNA) (c) a lipid molecule, in this case phosphatidal choline; (d) a segment of a polysaccharide.

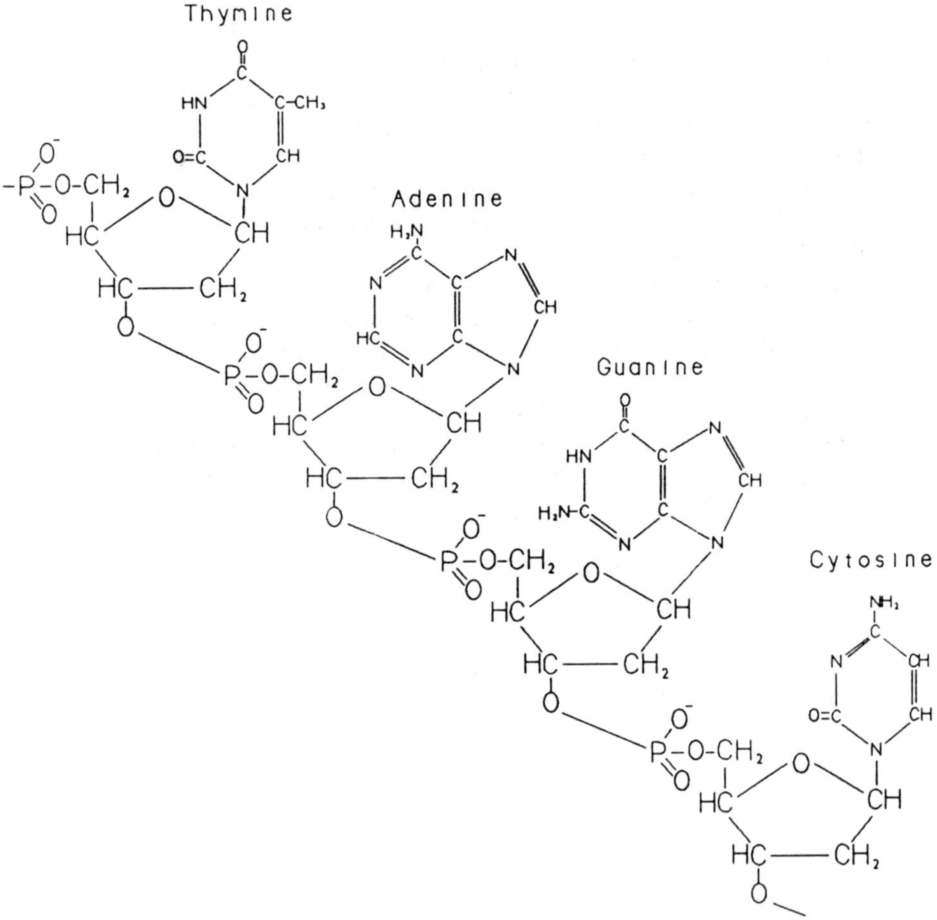

(b)

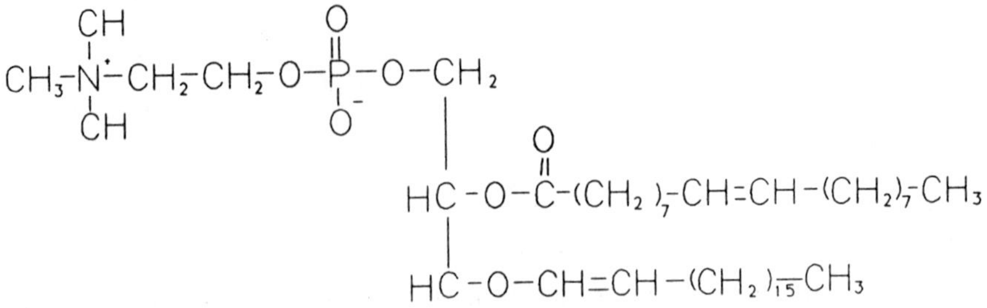

(c)

Figure 3.16 *(continued)*.

(d)

The distribution of biomolecules in cells, shown in Table 3.3 for an *E. coli* cell, is typical of most cellular species. Proteins are the most abundant components with nucleic acids, carbohydrates, and lipids following in that order. Carbon is common to all, while nitrogen and sulfur are critical constituents of the proteins and the nucleic acids, which themselves comprise about two thirds of the dry weights of the cells.

These four major biomolecular species also serve identical functions in all living cells. Nucleic acids store and transmit the genetic information that determines the structure and function of the proteins. Proteins serve, among many other biological functions, as cell membrane structural elements, as enzymes that catalyze the metabolic processes that govern the assembly of the cell's complex components, and as the cell's first line of defense in the form antibodies. Carbohydrates like cellulose function as extracellular structural elements, while in the form of polysaccharides like starch, they serve as energy sources for cellular activity. And the lipids are the major structural elements of the cell's boundary membranes as well as repositories of energy needs to maintain cellular activity.

In contrast to lipids and carbohydrates, which are constructed of repeating units of identical simple molecules, proteins and nucleic acids

Table 3.3 Bimolecular Components of *E. coli*

Biomolecule	Wt.%
Water	70
Proteins	15
Nucleic Acids	
DNA	1
RNA	6
Carbohydrates	3
Lipids	2
Other molecules & intermediates	2
Inorganic ions	1

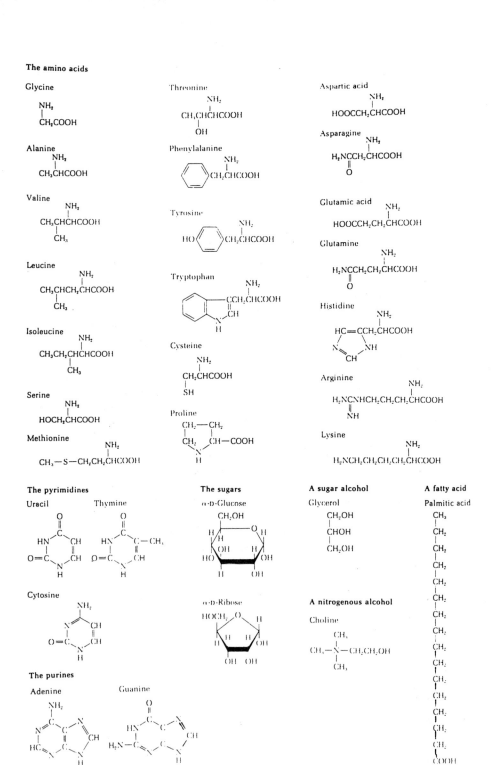

Figure 3.17 The thirty primordial biomolecules [Lehninger 1970, courtesy Worth Publishers].

are complicated structures of nonrepeating units, the sequences of which, in nucleic acids, comprise the genetic information necessary for cellular reconstruction and, in proteins, determine their structure and function. In lipids, the repeating units are usually methylene groups while many of the complex carbohydrates are built up from repeating sugar units. The nucleic acids, on the other hand, are constructed from sequences of four different nitrogenous chemicals called nucleotides, and proteins are chains of amino acids, twenty of which comprise the basic structural units for the proteins of all living matter.

In fact, the basic structural units of all living organisms are said to be the thirty "primordial biomolecules" shown in Figure 3.17 [Lehninger 1970, 21] and include the basic amino acids of the proteins; three pyrimidines and two purines, which are the nitrogenous bases upon which the nucleic acids DNA and RNA are built; and two sugars, two alcohols, and a fatty acid, which go into the structures of the carbohydrates and lipids. These thirty primordial biomolecules also take part in a host of other biosynthetic processes that produce over a thousand other biochemical moieties in the cell. The distribution of our subject elements within the primordial biomolecules and their biosynthetic products is about 18% carbon, 3% nitrogen, and 0.25% sulfur, with nitrogen and sulfur being critical constituents of the nucleic acids and the proteins.

Cells may be generally divided into two categories, autotrophs and heterotrophs. Autotrophic cells are able to use simple inorganic substances as energy sources while heterotrophic cells require preformed organic matter like glucose. Carbon utilizing autotrophs are essentially self-sufficient, using atmospheric carbon dioxide as their sole source of carbon. Heterotrophs obtain their carbon from other cells. Photosynthetic cells are autotrophic while most microorganisms and the cells of the higher animals are heterotrophic.

Carbon, nitrogen, and sulfur cycle among autotrophic and heterotrophic organisms as a consequence of their nutritional interdependence. The photosynthetic organisms produce reduced carbon compounds like glucose from atmospheric carbon dioxide; and they make use of soil nitrates and sulfates to produce organic nitrogen and sulfur compounds. Heterotrophs feed on the photosynthetic organisms as well as on one another to obtain their nutritional requirements. And all cells return carbon dioxide to the atmosphere and nitrogen and sulfur to the soil as products of their respiratory and metabolic cycles.

The Energetics of Cellular Metabolism

Clearly, for the chemical processes that produce the complex biomolecules of living organisms to proceed, energy must be expended. Sources of energy are large molecules like lipids, proteins, and polysaccharides that are already cellular constituents but may be, as in the case of heterotrophic organisms, ingested as food. Metabolic breakdown (or catabolism) of these molecules then releases the required energy.

A schematic of the catabolic process is shown in Figure 3.18. We will discuss the details of the scheme a little later. Suffice for now to mention that the energy provided by catabolic breakdown is conserved in the cells, to be used to produce new generations of biomolecules. This counterpart of catabolism, biomolecular production, is called anabolism.

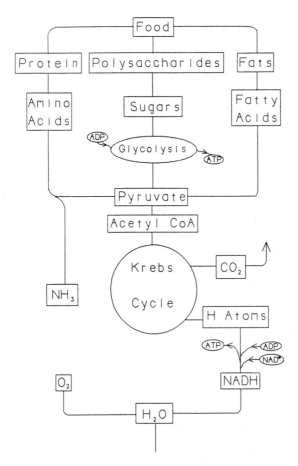

Figure 3.18 The catabolic process. Note that oxygen enters the process at the very end when it reacts with NADH to form water.

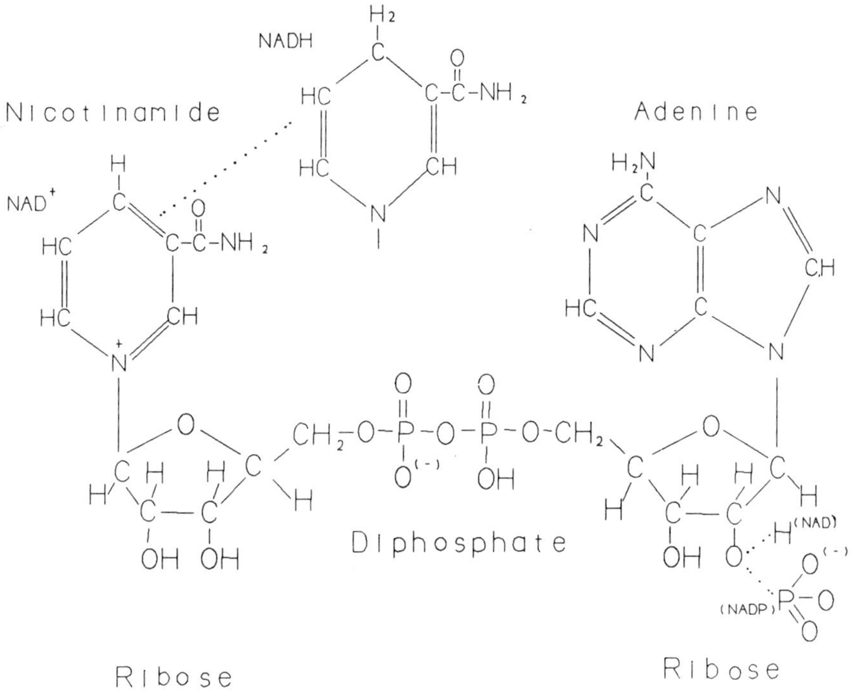

Figure 3.19   (a) An NAD$^+$ molecule. The structure of the reduced nicotinamide group is also shown as is the position of the phosphate addition to make NADP; (b) adensosine di- and triphosphate.

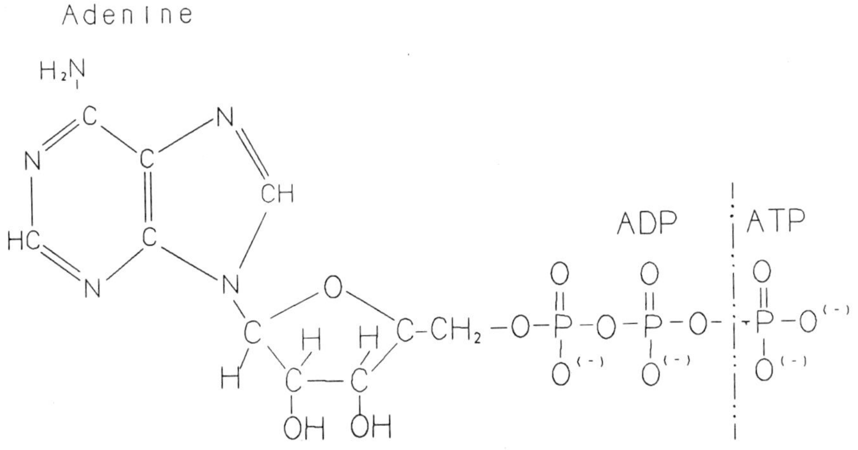

Energy conservation is accomplished by storing the energy released by catabolism in two systems of nitrogenous chemicals; these include the ADP/ATP and NAD^+/NADH couples. ADP is an acronym for adenosine diphosphate while ATP stands for adenosine triphosphate. NAD^+ is the oxidized form of nicotinamide adenine dinucleotide with NADH being the reduced form of the same compound. The structural formulas for these biochemicals are shown in Figure 3.19.

Returning to Figure 3.18, we note that polysaccharides are first degraded to simple sugars. The simple sugars then undergo a series of oxidation steps that ultimately yield carbon dioxide and release energy, which is stored in the ADP/ATP and NAD^+/NADH couples. The energy storage process in the ADP/ATP couple involves converting ADP to ATP, the latter of which is the higher energy moiety, eager to give up its extra phosphate group in other chemical reactions. And, as a consequence of the oxidation processes, NAD^+ is reduced to NADH, also the higher energy half of the couple, and eager to take part in other oxidation reactions. Proteins and fats are similarly degraded to simpler molecules like amino and fatty acids, and the released energy is also stored in the form of ATP and NADH, which ultimately provide the energy necessary to rebuild new biomolecules in the anabolic part of the metabolic cycle.

Photosynthesis

The interdependence of autotrophic and heterotrophic organisms begins with the photosynthetic production of reduced carbon from carbon dioxide. Photosynthesis is the mechanism that supports the growth of autotrophic organisms, and autotrophs are the sources of the carbon, nitrogen, and sulfur required to support the growth of the heterotrophs.

Photosynthesis in plants takes place in the plant cell organelles called chloroplasts. As shown in Figure 3.20, two categories of chemical reactions take place in the chloroplasts that ultimately lead to the production of reduced carbon compounds. The first, the "light dependent" reactions, produce ATP and a phosphorylated form of NADH designated NADPH (see Figure 3.19), the phosphorylated derivative coming from reactions between ATP and NAD^+. Solar energy, in concert with chlorophyll, which acts as a reaction catalyst, drives the synthesis of these compounds from ADP, NAD^+, and water, and produces oxygen as a by-product.

Following the light-dependent reactions, the "dark reactions" make use of the energy stored in the ATP and NADPH to drive the reaction shown in Figure 3.21, in which carbon dioxide reacts with a phosphorylated sugar called ribulose 1,5-bisphosphate to make 3-phosphoglycerate which, as is illustrated in Figure 3.22, is ultimately reduced to glyceral-

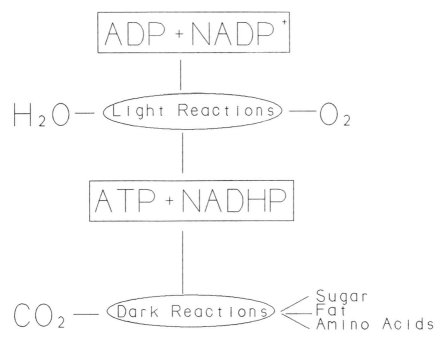

Figure 3.20 The dark and light reactions of photosynthesis.

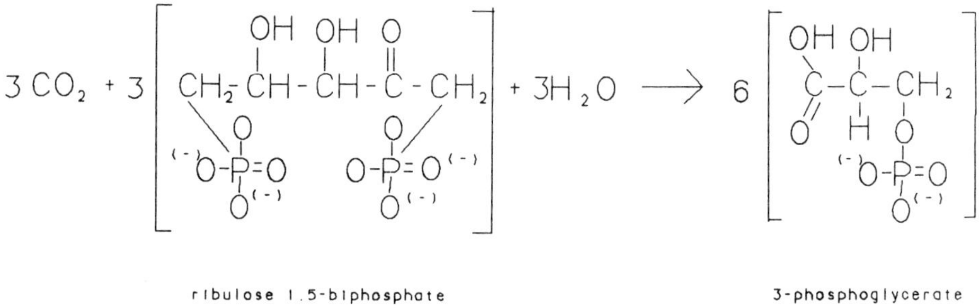

$$3\,CO_2 + 3 \left[\begin{array}{c} \text{ribulose 1,5-biphosphate} \end{array} \right] + 3H_2O \longrightarrow 6 \left[\begin{array}{c} \text{3-phosphoglycerate} \end{array} \right]$$

ribulose 1,5-biphosphate   3-phosphoglycerate

Figure 3.21 The initial dark reaction of photosynthesis.

dehyde 3-phosphate, the precursor to all other sugars, fats, and amino acids produced anabolically by the plant.

As can be seen in Figure 3.22, the reaction of CO_2 with the ribulose biphosphate is only the first step in a complex cycle of reactions that make continued use of ATP and NADPH to produce intermediates, which themselves ultimately regenerate the ribulose 1,5-biphosphate. The net result of the cycle is the single molecule of glyceraldehyde 3-phosphate that is produced at the expense of three molecules of CO_2, nine molecules of ATP, and six molecules of NADPH.

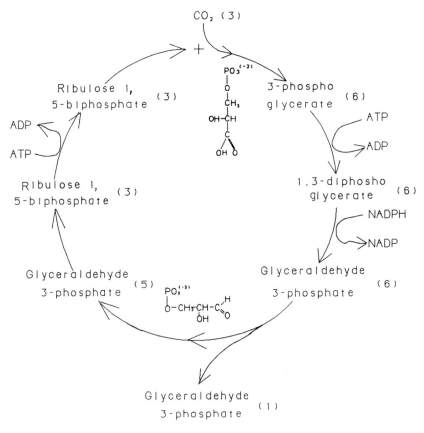

Figure 3.22   The dark reaction cycle of photosynthesis. The bracketed values denote the number of molecules engaging in the reaction progression.

Nitrogen Fixation and the Synthesis and Degradation of Amino Acids

Except for carbon and nitrogen, all the nutrients taken up by plants and transported through their vascular systems are present in the soil, derived from weathered rocks, decaying organic matter, or deposited by precipitation as is, for example, sulfur in the form of sulfate aerosols. Like carbon, all the nitrogen used by plants is ultimately derived from the atmosphere. But it is not used directly, as is carbon dioxide in photosynthesis. Instead, it may be assimilated from soil nitrogen, fixed there in the form of nitrates by soil bacteria or deposition of nitrogenous fertilizers, or it may be fixed directly from atmospheric nitrogen in the form of ammonia in the roots of leguminous plants as a consequence of symbiotic relationships with bacteria like the genus *Rhizobium*. It is then incorporated into photosynthesized carbon skeletons to produce cellular nitrogenous biomolecules like amino acids, ATP, or NADH.*

"*Symbiotic*" fixation involves bacterial infection of the root cells of the host plant. The bacteria then multiply in nodules attached to the plant roots. Stimulated by the plant, the bacteria produce an enzyme called nitrogenase, which catalyzes the conversion of atmospheric nitrogen to ammonia. The actual mechanisms involved in the conversion are not well understood, although it seems that nitrogenase behaves like a hydrogenase, an enzyme that can catalyze reduction reactions involving molecular hydrogen. It has also been observed that some nitrogen fixing bacteria produce an iron-containing protein called ferredoxin,[+] a strong reducing agent that is apparently necessary to the bacterium's ability to fix nitrogen. And, finally, it is known that ATP is consumed in the process. Thus, it appears that fixation may involve the reduction of hydrogen ions to molecular hydrogen, which then reacts with atmospheric nitrogen to form ammonia, all catalyzed by nitrogenase, and with the ATP providing the energy requirements to drive the reactions. It has been estimated that fixation of one mole of nitrogen by *Rhizobium* requires 25 to 30 ATP molecules [Alberts *et al* 1983, 1117].

All amino acids are derived, ultimately, from the ammonia fixed in plants. The primary bio-synthetic step is one involving glutamate dehydrogenase, an enzyme that catalyzes the synthesis of glutamic acid,

---

*Higher animals like man, for example, are capable of synthesizing only ten of the primordial amino acids required for protein synthesis. They must obtain the other ten, the "essential amino acids," from other organisms. Plants, however, are capable of synthesizing all twenty.
[+]An interesting sidenote is that ferredoxin is an exceptionally high sulfur-containing protein, being about 10% sulfur by weight.

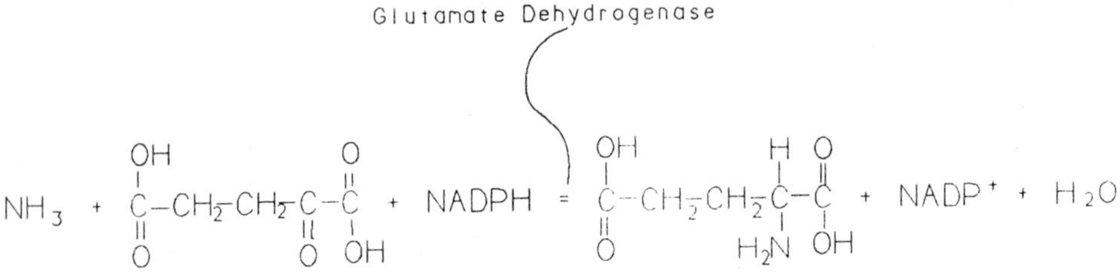

Glutamate Dehydrogenase

$$NH_3 \; + \; \underset{O}{\overset{OH}{C}}-CH_2-CH_2-\underset{OH}{\overset{O}{C}}-C \; + \; NADPH \; = \; \underset{O}{\overset{OH}{C}}-CH_2-CH_2-\underset{H_2N \; OH}{\overset{H \; O}{C}}-C \; + \; NADP^+ \; + \; H_2O$$

Figure 3.23   Nitrogen fixation, the synthesis of glutamic acid.

one of the thirty primordial biochemicals, from ammonia and glutaric acid. The reaction is shown in Figure 3.23.

Note that the α-keto glutaric acid in the reaction is one of a class of α-keto acids with the general formula

$$R-CO-COOH$$

In the reaction, the oxygen on the carbonyl group ajoining the carboxy group of the α-keto acid is substituted with an amine group from the ammonia to form the amino acid. The general formula for an amino acid is then

$$R-CHNH_2-COOH$$

The source of the α-ketoglutaric acid for this primary amino acid synthesis is in the Krebs cycle shown within the metabolic scheme of Figure 3.18, which will be discussed in more detail a little later.

Glutamic acid serves as the starting material for the synthesis of many other amino acids by transferring its amine group to other α-keto acid structures in transaminase reactions as shown in Figure 3.24. And, as a consequence of its position as the starting material for other amino acids, it also serves as the effective starting point in the synthesis of purines and pyrimidines; these are produced from secondary amino acids like glutamine, aspartic acid, and glycine, and are the nucleotide bases of the nucleic acids. The sequences of the nucleotide bases in the nucleic acids

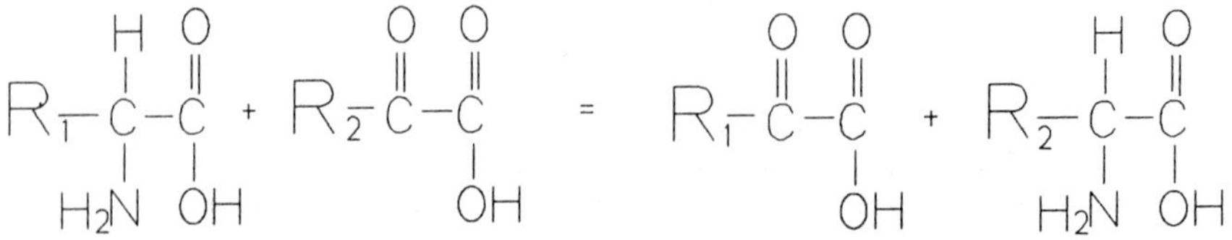

$$R_1-\underset{H_2N \; OH}{\overset{H \; O}{C}}-C \; + \; R_2-\underset{OH}{\overset{O \; O}{C}}-C \; = \; R_1-\underset{OH}{\overset{O \; O}{C}}-C \; + \; R_2-\underset{H_2N \; OH}{\overset{H \; O}{C}}-C$$

Figure 3.24   The transaminase reaction.

determine the assembly sequences of the amino acids in the synthesis of proteins.

Amino acids and other nitrogenous compounds may degrade via a route reverse to amino acid synthesis. The amino acids, or nitrogenous compounds degraded to amino acids, undergo a series of transaminase reactions that ultimately return to glutamic acid, which decomposes to ammonia and α-keto glutaric acid. A new synthesis cycle may then begin or the ammonia may be eliminated as waste.

Respiration

The respiration of all living organisms can be summarized by the chemical reaction

$$(CH_2O)_n + O_2 = H_2O + CO_2 + energy$$

where $(CH_2O)_n$ represents any organic moiety used by an organism as a source of energy. This simple reaction is just the reverse of photosynthesis and summarizes the cyclic transport of carbon dioxide within the biosphere. Like photosynthesis, however, the biochemical mechanisms involved in respiration are complex, probably because as organisms evolved, they adapted to the fact that the direct conversion of organic matter to water and carbon dioxide could yield energy only in the form of heat, which was generally unusable to them. So the catabolic scheme shown in Figure 3.18 evolved as a way to control the conversion of food to carbon dioxide and harness the energy released from the conversion as ATP and NADH. It is a noteworthy feature of the several stages of catabolism that oxygen does not enter the scheme until the very end and, as will be seen, that carbon dioxide is generated by cleaving it from other organic species rather than by direct reaction with oxygen.

Central to catabolism (and respiration) is the cycle of chemical events shown in Figures 3.18 and 3.25 that is variously called the Krebs cycle, the tricarboxylic acid cycle, and the citric acid cycle. It is in this metabolic process that carbon dioxide is generated from catabolized nutrients. The energy derived from its generation is then transferred in the form of hydrogen atoms to NAD^+ to make NADH; and in the reduction of NAD^+ to NADH, ADP is consumed to make ATP in a process called oxidative phosphorylation.

But note that catabolized nutrients do not enter the citric acid cycle directly. Rather, the form of entry is acetyl CoA, a condensation product of a purine based nucleotide called Coenzyme A (Figure 3.26) and acetic acid, the acetic acid being derived from the catabolized nutrients.

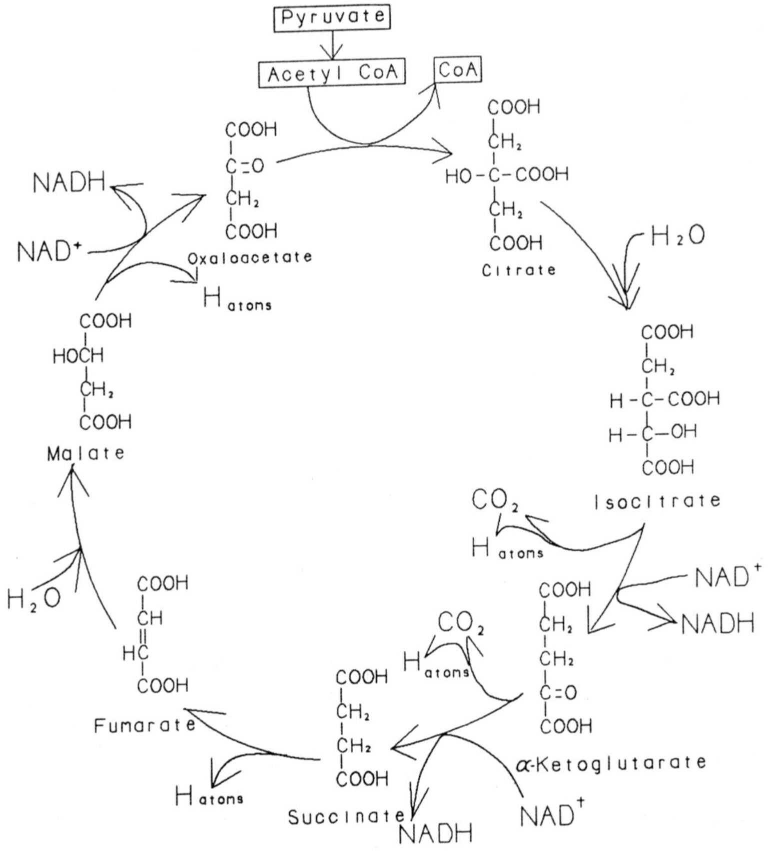

Figure 3.25 The Krebs or citric acid cycle. Note the formation of NADH and the production of carbon dioxide and hydrogen atoms.

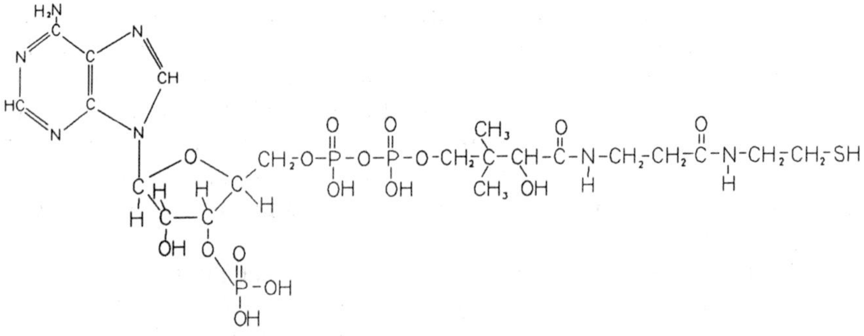

Figure 3.26 Coenzyme A. In the formation of acetyl CoA, the acetic acid binds to the sulfur atom of the sulfhydryl group.

The paths to acetyl CoA and entry to the citric acid cycle are complex, with carbohydrates, fats, and proteins each passing through different steps. Fats and carbohydrates enter as acetyl CoA. Some amino acids from catabolized proteins enter as acetyl CoA, but others degrade to intermediates that are compatible with entry at other points in the cycle. All catabolic products that are involved in respiration, however, inevitably pass through the citric acid cycle to be decomposed to carbon dioxide.

As is shown in Figure 3.18, carbohydrates first undergo glycolysis where they are degraded to pyruvic acid, an α-keto acid of the form $CH_3COCOOH$. As a result of a complicated series of multi-enzyme catalyzed reactions then, the pyruvic acid is oxidized to acetic acid and carbon dioxide, and the acetic acid is bound to the sulfhydryl group of Coenzyme A.

Several amino acid protein metabolites also follow the pyruvic acid path to acetyl CoA. Others degrade to different acids that condense with Coenzyme A and enter the cycle at points appropriate to the nature of the acid-CoA formed. And still others, as can be seen in Figure 3.27, enter directly by being degraded to intermediate products of the cycle. Thus, methionine, for example, after being degraded to succinic acid, enters the cycle as succinyl CoA. Glutamine is converted to glutamic acid and, after a transaminase reaction, enters the cycle directly α-keto glutaric acid.

Fats enter the Krebs cycle as acetyl CoA. They are first degraded to fatty acids, and the fatty acids enter the reaction cycle shown in Figure 3.28. After first binding to Coenzyme A, the fatty acids are shortened by two carbon atoms to make acetyl CoA. The residual acids then repeat turns through the cycle, each time losing two carbons to combine with additional Coenzyme A and form another molecule of acetyl CoA. The cycle is repeated until the whole of the original fatty acid is decomposed.

As can be seen in Figure 3.25, the acetyl CoA entering the Krebs cycle loses its bound acetyl group, which then reacts with oxaloacetic acid, a regenerating product of the cycle, to form citric acid, whence comes the cycle's name. In the next step, the citric acid is changed to isocitric acid, one CO_2 molecule is broken off, and one molecule of NADH is formed. And in the next step, that process is repeated, generating another molecule of CO_2 and another NADH. The remainder of the cycle then regenerates the oxaloacetic acid to start the cycle over again.

The respiratory cycle is finally completed with the reaction

$$2H^+ + 2NADH + O_2 = 2H_2O + 2NAD^+$$

in which the NADH produced in the citric acid cycle reacts with the hydrogen ions of cellular fluids and oxygen from the air to make water.

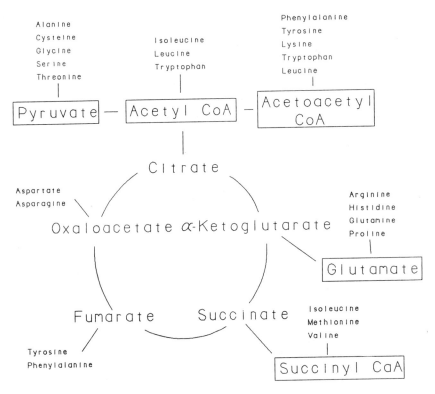

Figure 3.27   Amino acid entry points in the Krebs cycle.

Sulfur

Thus far, little has been said about sulfur, although it is present in all cells, primarily in proteins, but also in biomolecules like Coenzyme A and others that function as metabolic regulators. The principal repositories of sulfur in living organisms, however, are methionine and cysteine, the two sulfur containing amino acids.

Functionally, cysteine and methionine serve as components of proteins and as starting points for the biosynthesis of other biomolecules, sulfur-containing and otherwise. The terminal methyl group of methionine is particularly active because of its proximity to the molecule's sulfur atom (see Figure 3.17), and it is readily transferred to other biomolecules, making methionine a very active participant in general biomolecular synthesis. In addition, methionine in bacterial cells and in the mitochondria of eucaryotic cells serves as the chain initiating amino acid in the synthesis of the cell's proteins.

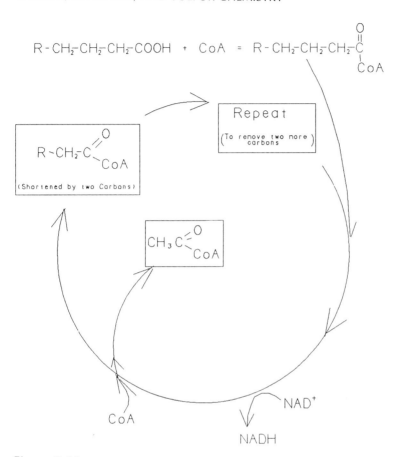

$$R - CH_2-CH_2-CH_2-COOH \quad + \quad CoA \quad = \quad R - CH_2-CH_2-CH_2-\overset{\overset{\displaystyle O}{\|}}{\underset{\underset{\displaystyle CoA}{|}}{C}}$$

Repeat

(To remove two more carbons)

$$R \text{~} CH_2-C\overset{\displaystyle O}{\diagdown_{CoA}}$$

(Shortened by two Carbons)

$$CH_3 C\overset{\displaystyle O}{\diagdown_{CoA}}$$

NAD^+

CoA

NADH

Figure 3.28   The degradation cycle of fatty acids to make acetyl CoA.

Cysteine plays a major role in the three dimensional structure of proteins. As can be seen in Figure 3.16, proteins are constructed of sequences of amino acids linked between the carboxy group of one and the amine group of another in a peptide bond. The peptide bonds constitute the backbone of proteins. But cysteines, interspersed among the protein's amino acids, can link to one another through their sulfhydryl groups (–SH) and form disulfide bridges between neighboring cysteine residuals. Thus, they can bind individual protein chains to one another or stabilize the complex three dimensional structures that these large molecules assume as they twist and fold back on themselves.

While plants and microorganisms can synthesize both cysteine and methionine, mammals can produce only cysteine, which they do in a complex series of biochemical steps starting with methionine supplied exogenously. Thus, methionine is one of the ten essential amino acids that mammals must obtain from food.

Sulfur passes through a number of oxidation stages in its biospheric cycling between organic sulfur and sulfates as is shown in Figure 3.29. Organic sulfur is produced by plants and microorganisms from inorganic sulfate by first forming the derivative phosphoadenosine-5-phosphosulfate (PAPS) shown in Figure 3.30. Further reduction to sulfite and sulfide is then catalyzed by the enzyme "PAPS reductase." Finally, catalyzed by the enzyme serine sulfhydrase, cysteine is synthesized from sulfide and serine according to

$$\underset{\underset{NH_2}{\mid}}{HOCH_2CHCOOH} + H_2S = \underset{\underset{NH_2}{\mid}}{HSCH_2CHCOOH} + H_2O$$

Methionine may then be synthesized by plants from cysteine via paths similar but reverse to those used by mammals to produce cysteine from methionine.

Return of sulfur to the environment is via hydrogen sulfide, which is a product of decay and putrefication processes. Bacterial degradation of proteins, for example, usually leads to release of products that have free sulfhydryl groups like cysteine. These products are then subject to decomposition by a class of enzymes called desulfhydrases found in soil and marine bacteria, the most common of which is the genus *Clostridium*.

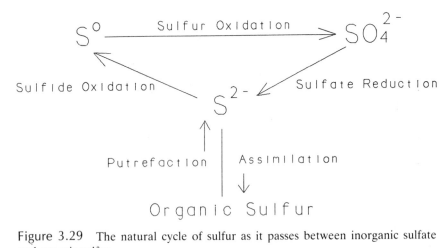

Figure 3.29 The natural cycle of sulfur as it passes between inorganic sulfate and organic sulfur.

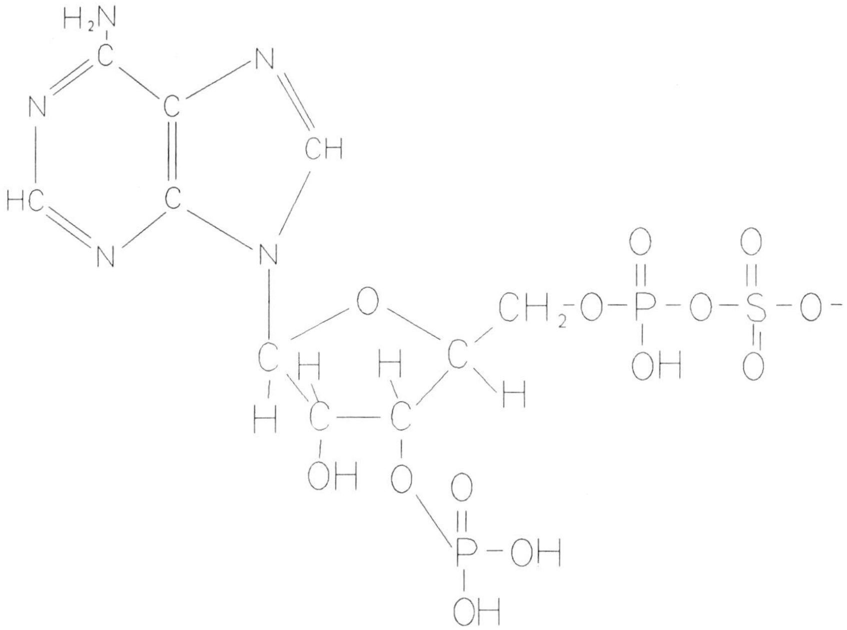

Figure 3.30   The structure of phosphoadenosine-5-phosphosulfate (PAPS).

Clostridium produces cysteine desulfhydrase, which catalyzes the removal of both amino and sulfhydryl groups from amino acids, producing pyruvic acid and hydrogen sulfide according to the reaction shown in Figure 3.31. The hydrogen sulfide produced is finally oxidized back to inorganic sulfate by a number of different classes of soil and marine bacteria, some of which can couple sulfide oxidation to carbon dioxide in a photosynthetic reaction that produces bacterial carbohydrate.

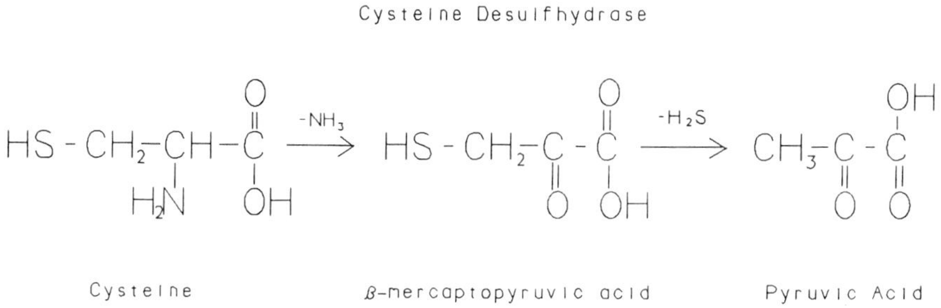

Figure 3.31   The release of ammonia and H_2S in the bacterial degradation of cysteine.

4

Carbon, Nitrogen, and Sulfur in Water Pollution

DEFINITION OF WATER POLLUTION

In earlier times, water was regarded as polluted when it was found to be unfit to drink. A more modern and more generalized definition of water pollution, however, considers not only potability but overall ecological health, because it is now understood that the ecological health of a water resource determines its ultimate usefulness to a modern society. Thus, a polluted waterway is one that is ecologically out of balance as a result of contamination.

An ecologically healthy aquatic system is self-cleaning. Through photosynthesis, large plants rooted to the bottom, and floating microscopic plants like algae and other phytoplankton, produce oxygen and provide food for consumers. Plant and oxygen consumers in the system are the zooplankton, or bottom dwelling microscopic animals and insects, fish, amphibians, mammals, etc. Additionally, there are other consumers called detritivores (detritus feeders) that live on organic debris at the water's bottom derived from the death, decay, and waste of all the organisms in the system. And finally, the decay microorganisms, the bacteria and fungi, recycle essential nutrients back to the oxygen and food producers.

The balance of such an aquatic ecosystem can be profoundly affected by a countless number of pollutants that may be added to it. And in the course of water's cycle, from evaporation from the ocean to its flow from the high mountains back to the sea, it may be used and reused by society with infinite potential for assimilating contaminant threats to its balance.

WATER POLLUTANTS AND THEIR ENTRY PATHS

The enormous number of waste contaminants that a modern industrial society produces defies cataloging. However, it is possible to categorize them into just a few classes which include:

1. Oxygen-Utilizing Wastes

These are biodegradable organics found in domestic sewage and industrial discharges. Oxygen starvation in water leads to the suffocation of marine organisms, anaerobic decay and, ultimately, to the death of the waterway. These wastes and their biodegraded products also contribute to increased levels of dissolved and suspended solids in water.

2. Synthetic Organic Compounds

This category includes detergents, pesticides, industrial solvents and chemicals, many of which are toxic to aquatic life as well as being oxygen utililizers and contributors to dissolved and suspended solid levels.

3. Plant Nutrients

Plant nutrients are nitrogen and phosphorous compounds derived from runoff from fertilized lands, seepage from fertilizer-contaminated ground water, leachants from animal wastes, and effluents from sewage treatment plants and domestic septic systems. Proliferation of nutrients leads to overabundance of marine organic matter and, ultimately, to waterway eutrophication.

4. Inorganic Chemicals and Minerals

These are discharged as waste products of industrial processes, drained from mining operations in the form of mineral acids, or deposited by acid rains. They are generally toxic to marine organisms.

5. Pathogenic Bacteria

Pathogenic bacteria enter water with domestic sewage and animal wastes washed from feedlots. They threaten water potability.

6. Sediments

Sediments are composed of particles of soil, sand, and minerals washed from the land by rain or irrigation water. They can smother a waterway's bottom life, fill reservoirs and harbors and, by surface adsorption, can transport other pollutants with them.

7. Radioactive Materials

Radioactive contaminants may be produced by mining or processing of radioactive ores, or be discharged from medical facilities, or be generated by nuclear weapons testing.

8. Heat

Heat generated by power and industrial plants may be discharged from them in the form of cooling waters. Increased water temperature resulting

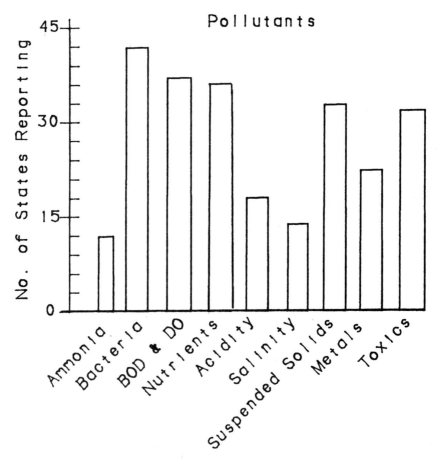

Figure 4.1   The distribution of water pollutants in the United States (from Environmental Quality—1984).

TABLE 4.1   Pollutant Categories and Sources

Categories	Sources
BOD/COD	Municipal waste treatment, pulp & paper mills, combined sewers, natural sources
Bacteria	Municipal waste treatment, combined sewers, septic systems, feedlots, urban runoff, pastures and rangeland, natural sources
Nutrients	Municipal waste treatment, agriculture, forest management, combined sewers, construction runoff, septic systems
Suspended solids	Agriculture, urban runoff, mining, construction runoff, combined sewers, forest management
Dissolved solids	Agriculture, mining, urban runoff, combined sewers
pH	Atmospheric deposition, mine drainage
Ammonia	Municipal waste treatment, combined sewers
Toxics	Industries, municipal waste, agricultural, land disposal of wastes, forest management, spills, combined sewers

from thermal pollution is detrimental to ecological balance because it accelerates chemical and biological activity.

Figure 4.1 illustrates the distribution of the most widely reported of these pollutant categories in the United States. Table 4.1 shows their possible sources [Environmental Quality—1984, 88]. They all may enter water resources in innumerable ways, but all entry paths can be distilled to the following: 1) direct discharge; 2) runoff and seepage; 3) transport across the air–water interface; and 4) transport across bottom sediment-water interfaces. Then, in dissolved and particulate form, they are dispersed and transported by currents, turbulent mixing, and by ingestion by migrating organisms.

WATER POLLUTION CRITERIA

Because of the vast number of possible water contaminants,* water pollution criteria, that is, the parameters used to generally characterize degrees of pollution, as well as discharges to waterways, are frequently nonspecific. Thus, color, odor, turbidity, temperature, and density may be used as physical criteria, while chemical criteria may include acidity,

---

*As of July 1985, EPA had approved about 1000 tests for about 250 inorganic, nonpesticide organic, and pesticide organic pollutants [CFR 1985, 248]. The list includes tests that characterize water pollution generally as well as tests for substances that have been recognized as particularly threatening.

alkalinity, salinity, oxidation-reduction potential, dissolved oxygen (DO), total carbon (TC), organic carbon (TOC), and oxidizable matter. Depending upon need, abnormal levels in any of these criteria may then precipitate procedures to identify and quantitate more specific contaminants.

The meaning and significance of the physical parameters should be obvious. Several of the chemical criteria, however, deserve some additional discussion.

Acidity is a measure of the hydrogen ion concentration of the water. It is usually defined in terms of pH where

$$pH = -\log[H^+]$$

and $[H^+]$ represents the hydrogen ion concentration. pH is normally determined electrochemically,* by measuring the difference in electrical potential between each of a pair of electrodes immersed in the water, one of which is a special glass electrode sensitive to hydrogen ions while the other is a so-called standard electrode. pH may also be determined by use of papers impregnated with pH sensitive dyes.

Alkalinity is a measure of the water's ability to neutralize incoming acid. The capacity to neutralize acids is dependent upon the concentrations of dissolved basic substances like carbonates, bicarbonates, ammonia, phosphates, and silicates.

Salinity may be determined by measuring the electrical conductivity of the water, which is dependent upon the concentration of dissolved ionic salts. Determination of total dissolved solids in the water is also related to salinity, but it includes dissolved organics as well.

Oxidation-reduction potential (ORP) is, like pH, determined electrochemically. It is reported in terms of the difference in potential between a metal electrode and a standard electrode and is designated pE. ORP is, in a way, is a measure of the aerobic condition of the water. That is, it is an indicator of the propensity of the constituents of the water to be oxidized or reduced. Thus, as is illustrated in Table 4.2, at low values of ORP, products generated by chemical and bacterial action in the water will be in reduced form, while at high values of ORP, they will be in oxidized form.

Oxidizables are commonly expressed in two ways, as biochemical oxygen demand (BOD) and as chemical oxygen demand (COD). Biochemical oxygen demand is a measure of the amount of biodegradable

---

*Electrochemical measurements are discussed in more detail in a later chapter.

TABLE 4.2   Product Dependence on Oxidation Reduction
Potential

Element	High-pE products	Low-pE products
Carbon	CO_2, CO_3^{2-}	CH_4
Nitrogen	NO_3^-	NH_3, NH_4^+
Sulfur	SO_4^{2-}	H_2S, HS^-

organic matter in water. It is the total amount of oxygen consumed
when a sample, initially saturated with oxygen, is subjected to bacterial
decomposition at a fixed temperature for a period of five days. Chemical
oxygen demand is the amount of strong oxidizing agent* that is consumed
by a fixed quantity of water. Chemical oxygen demand is a measure of
all oxidizable substances in the sample including inorganics like chlorides,
sulfides, and ammonia.

Both BOD and COD are tests for relatively easily oxidized substances.
They were designed primarily as quality criteria for industrial and municipal wastes. They do not measure aromatics or aliphatic carbon compounds, for example. Thus, in situations where those kinds of pollutants

TABLE 4.3   BOD Generation by Industry Type

Industry type	Waste water generation (billions of gallons)	BOD (millions of tons)
Animal feed	690	4300
Textiles	140	890
Paper	1900	5900
Chemicals	3700	9700
Petroleum and coal	1300	500
Rubber and plastics	160	40
Primary metals	4300	480
Machinery	50	60
Electrical machinery	91	70
Transportation equipment	240	120
Other manufacturing	450	390
Domestic waste treatment[a]	5300	7300

[a]Servicing 120 million people and assuming 120 gal/person/day of waste water and 1/6 lb/per-
son/day of BOD.

---

*Usually, potassium dichromate ($K_2Cr_2O_7$) is used.

are expected, measurements of total carbon or total organic carbon are preferred. The utility of BOD as an indicator of potential industrial and municipal sewage pollution is illustrated by the data in Table 4.3, in which typical BOD discharges are shown for several industries [Horne 1978, 674].

WATER AS A REPOSITORY FOR POLLUTANTS

Water Chemistry

It would appear that water serves as a welcoming receiver of contaminants of all kinds. Its behavior in that regard derives from its somewhat unusual chemical and physical characteristics — unusual, that is, relative to substances that are chemically similar to it like hydrogen sulfide, methane, and ammonia. These latter three are the hydrides of sulfur, carbon, and nitrogen, respectively, while water is the hydride of oxygen. Since carbon, nitrogen, and sulfur are all close neighbors to oxygen in the periodic chart, one might expect their hydrides to share similar chemical and physical characteristics, but they don't in several important respects.

Water is a liquid at room temperature and its solid form, ice, is less dense than the liquid. These are, perhaps, water's two most unusual features compared to H_2S, CH_4, and NH_3. It is also a dissociating solvent, i.e., ionic substances tend to dissociate in it. Liquid ammonia shares this property to a limited extent but hydrogen sulfide and methane do not. Water is insoluble in non-polar organic solvents and exhibits only limited solvent activity with non-polar compounds, but it is a good solvent for polar solutes. Hydrogen sulfide and methane, contrariwise, are soluble in non-polar solvents, and their liquid phases exhibit only limited solubility for polar solutes. These differences from closely related compounds result from strong hydrogen bonds between water molecules.

Hydrogen bonding, as we have mentioned before, is the attraction of bonded hydrogens to electronegative atoms. In water, hydrogen bonds form because the bonding electrons of its two hydrogens are strongly polarized toward the oxygen, imparting electropositive character to them. As a consequence, they are able to associate with the oxygens of other water molecules. Of the elements mentioned above, carbon, nitrogen, sulfur, and oxygen, oxygen is the most electronegative (Table 3.1) and forms the stongest hydrogen bonds, making water the most highly associated of all the hydrides mentioned with the highest melting and boiling points.

The orientation of the hydrogen atoms in the water molecule is shown in Figure 4.2a. As can be seen, they reside at the corners of a distorted

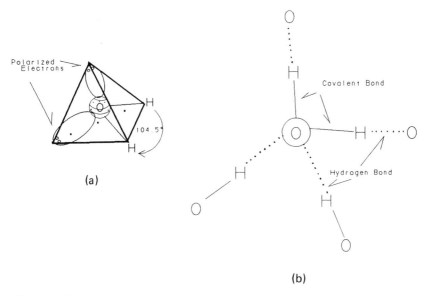

Figure 4.2   (a) The structure of the water molecule. (b) Hydrogen bonding in ice and ice clusters.

tetrahedron, the angle between them being 104.5°. The remaining bonding electrons of the oxygen tend to concentrate at the other two corners of the tetrahedron, and it is at those corners that hydrogen bonds associate water molecules into configurations of the form shown in Figure 4.2b.

These configuations are fully formed in ice. In the liquid, however, they exist as regions of open structured ice-like clusters mixed with more densely packed non-hydrogen bonded individual water molecules. This is the characteristic of the liquid that leads to its higher-than-ice density at the freezing point and to its density maximum at 4°C.

With increasing temperature, the ice clusters melt, contributing to a decrease in the volume of the mixture. But simultaneously, the increased kinetic energy of the mixture drives it to expand. The two actions counter one another and at 4°, a maximum density point occurs. Below maximum density, formation of structured clusters predominates, lowering density as temperature decreases so that the fully formed hydrogen bonded struc-ture of ice is less dense than the liquid. Above maximum density, forma-tion of non-structured regions begins to predominate and the mixture behaves normally, expanding with increasing temperature.

The tetrahedral structure of the water molecule, as well as its highly polar oxygen-hydrogen bonds, provide it with its unusual solvent characteristics. As a highly associated liquid, it would not be expected to be a very good solvent, because solvent-solvent interactions would be expected to be stronger than any solvent-solute interactions that could occur. In fact, except at its surface, it is not a good solvent for non-polar or weakly polar solutes. But it is an excellent solvent for polar solutes, a property deriving from the water molecule's shape and polarity, which provide it it with an extraordinary ability to reduce the force of attraction between electrically charged particles, i.e., provide it with an extraordinarily high dielectric constant.

The dielectric constant of a material is a measure of its ability to separate two electrical charges, the force between the two being given by Coulomb's law

$$F = \frac{q^+ q^-}{4 \pi r^2 \epsilon}$$

where q^+ and q^- are the charges, r is the distance between them, and ϵ is the dielectric constant of the medium in which they reside. Thus, large values of ϵ reduces the force between the charges. Water, as a high dielectric medium, therefore, works to reduce the attractive forces between polarized pairs and allows them to separate and dissolve.

Though the highly polar character of water reduces its bulk solvent power for non-polar solutes, their solution does occur at water's surface, i.e., at the air-water interface, where the polarity of the water is modified by its proximity to the non-polar gases of the air. In fact, non-polar and partially polar molecules, like long chain fatty acids, concentrate at the air-water interface. This concentration effect is large, organic pollutant concentrations at water's surface having been observed to be as large as 10^4 that of the bulk [Mullins 1977, 390]. Not only does the concentrated surface film make sampling procedures for pollution monitoring purposes more difficult to implement, it affects material transport processes between the air and the bulk of the water. Thus, it may inhibit oxygen replenishment from the atmosphere as well as photosynthesis by marine plants, and it can shift the carbon dioxide-water equilibria.

Water's Response to Pollution

Perhaps the most important criterion of water quality is its level of dissolved oxygen,* which can be affected by four processes: photosynthesis, respiration, oxygen transport, and waste oxidation. In non-polluted water, there is a normal diurnal cycle in DO level that is dependent upon photo-

---

*The minimum level of DO for fish to survive, for example, is about 3 mg/l.

synthesis, respiration, and aeration by atmospheric oxygen. During daylight hours, photosynthetic processes consume carbon dioxide and produce oxygen. If oxygen saturation occurs as a result, the excess is consumed by venting to the atmosphere. During the night, oxygen is added via transport from the atmosphere and, at the same time, is consumed by respirating plant matter. The net result is the cycle shown in Figure 4.3.

Entry of high levels of oxidizable organic matter, however, can disrupt the normal diurnal cycle in a waterway, because DO may be rapidly depleted as the organic matter decomposes. Countering the depletion, oxygen may be transported from the atmosphere and replenish the DO level at a rate proportional to the extent to which it is less than its saturation value. However, if replenishment cannot keep up with the organic challenge, a condition known as "oxygen sag" occurs. That is, the normal minimum in DO of the diurnal cycle is lowered and extended in time. Ultimately, as the organic matter is consumed, the rate of re-oxygenation can exceed the DO consumption rate, and the levels can return to normal. The profile of oxygen sag as a function of distance from

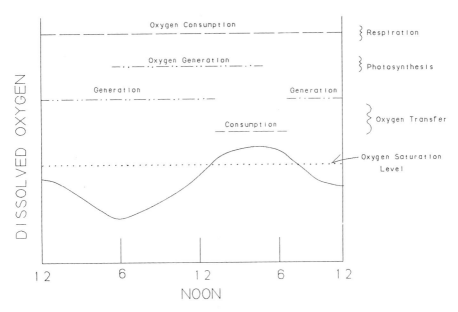

Figure 4.3   The diurnal variation of dissolved oxygen. Dissolved oxygen is always consumed by respiration and generated by photosynthesis. It is also generated by oxygen transfer from the atmosphere and consumed by venting to the atmosphere when its levels exceed the water's saturation value.

the point of entry for a moving stream, or of time in still waters, represents the normal self-purification ability of the waterway.

Chemically, the processes underway during self-purification may be expressed as

$$(CH_2O)_n \cdot (NH_3)_m + O_2 + \text{aerobic organisms} = xCO_2 + yH_2O + zHNO_3$$

The oxidizable organic matter in this reaction is characterized as $(CH_2O)_n \cdot (NH_3)_m$ [Horne 1969, 208], and the oxygen is that provided by the water. If the challenge to the water exceeds its oxygen based self-purification capacity, however, the water is said to become anoxic, and other oxidizing mechanisms come into play.

First, the nitric acid produced by the aerobic oxidation is utilized by denitrifying bacteria

$$(CH_2O)_n \cdot (NH_3)_m + uHNO_3 = xCO_2 + yH_2O + zN_2$$

and when all the nitric acid is consumed,* bacterial sulfate reduction occurs

$$(CH_2O)_n \cdot (NH_3)_m + uSO_4^{2-} = xCO_2 + yH_2O + zH_2S + qNH_3$$

producing the hydrogen sulfide that is associated with polluted waters.

Atmospheric Gases in Water

All atmospheric gases are found in water, their concentrations being dependent upon temperature and salinity, with low temperatures and salinities favoring higher concentrations. Oxygen solubility ranges between about 4 ml/l in sea water to about 10 ml/l in fresh water at 0°C and atmospheric pressure. Carbon dioxide, on the other hand, because it reacts with water to form carbonic acid, is about 35 times more soluble than oxygen on a weight basis. Nitrogen's solubility is about half that of oxygen's. Sulfur dioxide, like carbon dioxide, reacts with water. Its solubility is quite high, with saturated solutions containing about 53% by weight of the gas.

At equilibrium, the pressure exerted by a dissolved gas is equal to its partial pressure in the atmosphere,† and the rate of exchange of molecules

---

*Intermediate products of denitrification may include nitrous oxide and nitrites as well.

†Dalton's Law of Partial Pressures states that the total pressure exerted by a mixture of gases is equal to the sum of the partial pressures of each component, with the partial pressure of each being proportional to its concentration in the mixture.

between the two media is equal. When the system is not at equilibrium, i.e., the solution is either supersaturated or undersaturated, there is a net transfer of gas from one medium to the other, and the rate at which the transfer occurs depends upon the rate at which gas molecules can diffuse through the surface film at the air–water interface. Since the nature and thickness of the surface film are dependent upon factors like turbulence and chemistry, the transfer rate is highly variable, but in all cases affects the rate at which challenges to the water system's overall gas balance can be compensated. Thus, the oxygen sag curve, for example, will be affected by the surface film.

CARBON, NITROGEN, AND SULFUR WATER POLLUTANTS

Carbon and the Carbon Dioxide—Carbonate Equilibrium

Carbon Pollutants

Carbon pollutants can enter water in countless forms, some of which we have already discussed. Non-biodegradable carbon, or at least carbon in forms that have very long residence times, include halogenated pesticides like DDT, aromatic solvents, petroleum products, and detergents like alkyl benzene sulfonates. Biodegradable carbon pollutants have relatively short residence times. They are mostly organic wastes derived from sewage, industrial discharges, and agricultural runoff.

The long-term effects of many of the nonbiodegradable carbon pollutants have been well publicized and will not be discussed here, except to note that their effects on the ecological balance of water are nearly as numerous as they are, covering a range from outright toxicity to formation of foams that can inhibit oxygen and carbon dioxide transport across the air-water interface. Biodegradables, as we have mentioned, challenge the DO levels of their return sites as well as provide nutrients for excessive growth of aquatic organisms.

Regardless of the type of carbonaceous matter entering water, however, the end product of its decomposition, whether aerobic or anaerobic, is carbon dioxide. Thus, the paths of entry of carbon dioxide into water are three-fold: via the decomposition of carbonaceous matter; via transfer from the atmosphere across the air-water interface; and via the respiration of aquatic organisms.

The Carbonate Equilibrium

The reaction of carbon dioxide with water, which forms carbonic acid, bicarbonates, and carbonates has already been described. What was not emphasized in that description was the importance of the equilibria involved, which are among the most important of the chemical systems of the biosphere. Those equilibria affect the net transfer of CO_2 between air and water, the water's acidity and, thus, its overall chemistry, the growth of marine organisms, and the deposition of carbonaceous sediments.

The solution of carbon dioxide in water can be described by the equation

$$CO_2 + H_2O \overset{K_1}{=} H^+ + HCO_3^- \overset{K_2}{=} H^+ + CO_3^{2-}$$

where K_1 and K_2 are characteristic equilibrium constants defined as

$$K_1 = \frac{[H^+][HCO_3^-]}{[CO_2]}$$

and

$$K_2 = \frac{[H^+][CO_3^{2-}]}{[HCO_3^-]}$$

with the brackets signifying concentration levels.

And, since both of these equilibria must be satisfied simultaneously, the equilibrium constant equations can be solved in terms of the pH of the water, i.e.,

$$pH = -\log[H^+] = (1/2) \log \frac{[CO_3^{2-}]}{K_1 K_2 [CO_2]}$$

As can be seen, at high acidity levels (low pH), higher concentrations of carbon dioxide are favored and at low acidity levels (high pH), higher concentrations of carbonate ion are favored. At low pH, then, the system will tend to vent carbon dioxide to the atmosphere while at higher values, it will tend to absorb it.

Furthermore, rearranging the equilibrium constant equations in terms of the hydrogen ion concentration gives

$$[H^+] = \frac{K_1[CO_2]}{[HCO_3^-]} \text{ and } [H^+] = \frac{K_2[HCO_3^-]}{[CO_3^{2-}]}$$

Thus, high levels of bicarbonate have complementary effects on the hydrogen ion concentration. That is, they elevate it to satisfy one equilibrium and decrease it to satisfy the other, the net effect being that high levels of bicarbonate keep the hydrogen ion concentration at an intemediate level.

One may also view the equilibria contrariwise—as controlling the acidity of the water. That is, if processes underway in the system tend to maintain high levels of bicarbonate, the pH of the water will remain in the neutral region. Intrusions of high concentrations of carbon dioxide, on the other hand, as a result of rapidly decomposing organic matter for example, will lower the pH, while the dissolution of carbonate sediments will drive the system to higher pH levels. These effects are illustrated in Figure 4.4, which shows the concentration distribution of carbon dioxide, bicarbonate, and carbonate ions as a function of pH.

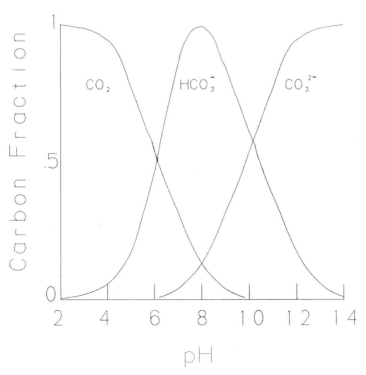

Figure 4.4 The carbonate equilibrium showing species concentration as a function of pH (from Fergusson 1982).

Nitrogen Compounds

Fixation

The ultimate source of nitrogen in the hydrosphere is atmospheric nitrogen that passes through the surface film at the air-water interface and is then fixed by the action of blue green algae. The initial product of fixation is ammonia which, depending upon the pH of the receiving water, may be converted to ammonium ion. The ammonia (or ammonium) is then converted to nitrogen oxides which are assimilated by resident aquatic organisms to make amino acids and proteins. Ammonia is also a product of the bacterial decomposition of waste organic matter, that process being part of the nitrogen cycle in water in which consumers return nitrogen to the environment. In an ecologically balanced water system, all the processes maintain a level of nitrogen adequate to provide the nutrient requirements of all the resident organisms.

Nitrification

The ammonia produced by fixation is oxidized (nitrified) by two species of autotrophic bacteria, the Nitroso group of the type *nitrosomonas* and the Nitro group of the type *nitrospira*. Thus,

$$nitrosomonas$$
$$4NH_4^+ + 7O_2 = 4H^+ + 4NO_2^- + 6H_2O$$

and

$$nitrospira$$
$$2NO_2^- + O_2 = 2NO_3^-$$

The conversion process is, of course, sensitive to pollutant intrusions. So, for example, excess ammonia from fertilizer runoff has been found to stimulate overpopulation of *nitrosomonas*, resulting in the production of excess nitrites [Mullins 1977, 380].

And changes in acidity influence the bacterial process which, as can be seen is, itself, an acid generator. Thus, any agent present in the water that can neutralize acidity will also help to stimulate the ammonia-to-nitrite conversion, while acid intrusions will inhibit it. It has been found, for example, that fertilizers made from salts of strong acids like ammonium sulfate are nitrified more slowly in acid soils than are alkaline fertilizers like anhydrous ammonia and urea [Hauk 1984, I–C, 114]. In alkaline soils, on the other hand, there seems to be no difference in the nitrification

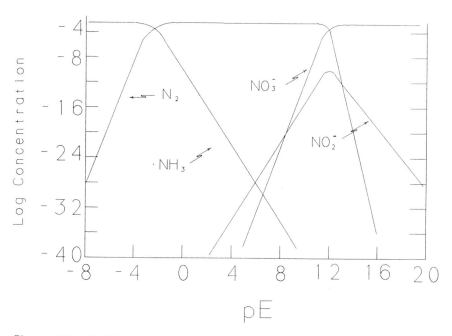

Figure 4.5   Equilibrium forms of nitrogen at pH 7 as a function of ORP (from Horne 1978).

rate, implying that the natural acid neutralizing capacity of the alkaline soil overcomes the influence of fertilizer type on the bacterial action. The acid-alkaline character of water similarly affects bacterial nitrification processes.

Nitrification is also influenced by the oxidation-reduction potential of water. Thus, the oxidation state of the resident nitrogen varies as a function of pE as is shown in Figure 4.5 [Horne 1978, 264]. As can be seen, at low values of pE, ammonia is the predominant species. At elevated values, nitrites and nitrates predominate.

Denitrification

Denitrification is the bacterial reduction of nitrites and nitrates to elemental nitrogen, which may be returned to the atmosphere or remain in solution to be reused in fixation. It is thought to be a step-wise reduction process of the type

$$NO_3^- \rightarrow NO_2^- \rightarrow NO \rightarrow N_2O \rightarrow N_2$$

with net reactions of the kind

$$4CH_2O + 4NO_3^- = 4HCO_3^- + 2N_2O + 2H_2O$$

and

$$5CH_2O + 4NO_3^- = H_2CO_3 + 4HCO_3 + 2N_2 + 2H_2O$$

underway.

Denitification bacteria are usually of the class of so-called chemoheterotrophs, organisms that make use of carbonacous materials as their sole source of energy and growth. They are, however, unique in that they can use either oxygen or oxidized nitrogen compounds for respiration [Hauk 1984, 116]. Under conditions of normal levels of dissolved oxygen, they behave like aerobic bacteria, using oxygen. At reduced oxygen levels, however, they consume nitrates and nitrites.

Denitrification, we note, produces nitrous oxide and is thought to be one of the processes responsible for the net production of that gas by ocean waters. However, there is some debate underway, since denitrification occurs in water only under anoxic conditions, and elevated levels of nitrous oxide have been observed over ocean waters containing levels of DO too high to allow for anoxia [Hauk 1984, 118].

We have already noted that the sources of nitrogen intrusions that can overload the normal nitrogen cycle of water include municipal sewage, industrial wastes, fertilizer and animal waste leachants and runoff, and acid rain.

Sulfur and Its Compounds in Water

Sulfur, as sulfate, is the fourth most abundant element in sea water, being about 1/20 the concentration of chloride. It is also present in fresh water, comprising, on average, about 12% of the weight of dissolved salts found there [Horne 1978, 247]. And many mineral springs contain sulfur in the form of sulfides at levels of the order of several grams per liter. However, despite the high *natural* concentrations of sulfur in the hydrosphere, that element still ranks among the most profuse of the environmental pollutants.

Sulfur Pollutants

The source of much of the environmental proliferation of sulfur is sulfuric acid, which is, perhaps, the most commonly used industrial chemical. In fact, it is so universally used that its consumption sometimes serves as an index of the economic health of a nation. Ultimately, a goodly percentage of that sulfur consumption finds its way into waste production and discharge as a pollutant.

Agricultural applications comprise the largest percentage of the world's sulfur consumption. Thus, about 40% of the production of sulfur (about 27 million tons in 1967) is used to make fertilizers—superphosphate and ammonium sulfate—and another 20% is used to manufacture insecticides and fungicides. Superphosphate ($Ca(H_2PO_4)_2$) is the solubilized form of the naturally occuring mineral apatite ($Ca_5(PO_4)_3OH$), converted thereto by sulfuric acid. Ammonium sulfate is prepared by reaction of sulfuric acid with ammonia. The agricultural applications (fertilizers, insecticides, and fungicides), of course, provide an immediate path to water pollution through runoff.

Another major source of sulfur as a water pollutant is the pulp and paper industry. In the USA in 1971, about 45 million tons of pulp were processed and another 13 million tons of paper materials were recycled. Seventy percent of the total amount of sodium sulfate consumed in the country went into those activities [Fergusson 1982, 180].

Sulfur in the form of sodium bisulfite is used in paper manufacture to solubilize lignins. Lignins are aromatic polymeric materials associated with wood cellulose and serve as a kind of glue in wood, holding the cellulose fibers together. To manufacture quality paper, they must be removed. Treating wood pulp with sodium bisulfite produces sulfonic acid groups on the lignin's aromatic rings and solubilizes them.

Typically, the waste liquor from a paper mill can contain about 100 g/l of solids of which about 6–8% is sulfur and sulfur dioxide. In other terms, production of a ton of dry paper can produce 600 kg of lignin, 200 kg of SO_2 combined with lignin, 90 kg of CaO combined with lignin sulfonic acid, and 370 kg of other organic waste [Horne 1978, 676]. This quantity of waste is an enormous oxygen depleting load and, in addition, its sulfur concentration represents a serious toxic threat to receiving waters.

And still another source of sulfur pollution is acid mine drainage, a high sulfuric acid containing effluent from mining operations. Acid mine drainage results from mining activities that expose sulfur-bearing minerals like pyrite (FeS_2) to the atmosphere or to sulfur-using bacteria like *thiobacillus*. Weathering of the pyrite converts it to ferric hydroxide and sulfuric acid

$$4FeS_2 + 15O_2 + 14H_2O = 4Fe(OH)_3 + 8H_2SO_4$$

and the *thiobacillus* use the pyrite as an energy source, converting it to ferrrous sulfate and sulfuric acid according to

$$2FeS_2 + 2H_2O + 7O_2 = 2FeSO_4 + 2H_2SO_4$$

It has been estimated that about 8 million tons of sulfuric acid are thus produced annually as a result of coal mining operations in the US alone [Fergusson 1982, 52] and, unless controlled, these acid wastes wash readily into neighboring waterways.

The Equilibrium of Sulfur Compounds in Aqueous Solution

The predominant sulfur species in natural water is sulfate. However, depending upon the water's acidity and ORP, it may exist in stable form as bisulfate (HSO_4^-), hydrosulfide (HS^-), and dissolved hydrogen sulfide. Hydrogen sulfide also, as we have mentioned before, dissociates to a small extent according to

$$H_2S = H^+ + HS^- = 2H^+ + S^{2-}$$

with the reaction moving to the right as pH increases. The insolubility of mineral sulfides like FeS_2, however, generally precipitates sulfide ions out of natural water solutions when they form.

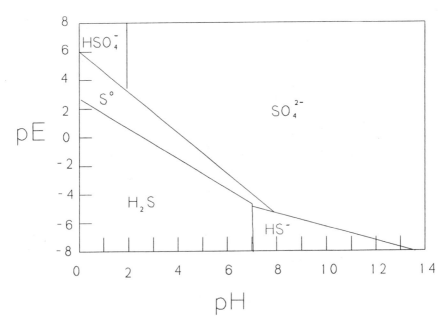

Figure 4.6 Dependence of the equilibrium forms of sulfur on pH and ORP (from Chen 1974).

Elemental sulfur is stable in acid waters only, existing, generally, as an insoluble colloidal suspension. In alkaline solution (pH > 8), however, it reacts with hydroxide and, depending upon the ORP, may form hydrosulfides, thiosulfates, and sulfates according to

$$4S + 4OH^- = 2HS^- + S_2O_3^{2-} + H_2O$$

and

$$4S + 5OH^- = SO_4^{2-} + 3HS^- + H_2O$$

with the thiosulfates and hydrosulfides converted to sulfates under conditions of high oxidation potential.

Although other species like sulfite, bisulfite, polysulfides, etc., may exist in water, they tend to be thermodynamically unstable in the natural water environment and quickly convert to these other forms [Chen 1974, 109]. The pH–pE dependence of the various forms of stable sulfur in water is shown in Figure 4.6.

SEWAGE TREATMENT

Because it is one of the more prominent sources of water pollution, and a significant generator of pollutants composed of our subject elements, the treatment of sewage deserves special attention.

Modern sewage treatment consists, generally, of at least two stages. These include primary treatment, basically a straining process that removes bulk solids, and secondary treatment, which processes the effluent from the primary treatment stage to remove biodegradable organic matter. In selected situations, a third or tertiary treatment step may be included, which removes substances that remain in the effluent from secondary treatment like residual BOD, non-biodegradable organics, and plant nutrients.

In primary treatment, raw sewage is screened to remove gross solids and then passed to settling tanks where suspended solids are allowed to sediment into sludge pools. The effluent liquids from the primary settling tanks are then subjected to one of several secondary treatment techniques, either activated sludge treatment, trickling filter treatment, or oxidation pond treatment.

Activated sludge treatment is a process in which primary effluent is passed to an aeration tank, where it is stirred aerobically, together with sludge laden with bacteria capable of decomposing its organic content. After a period of time, the aerobically treated secondary sewage (called the mixed liquor) is passed to a settling tank where remaining solids are allowed to separate by subsidence. These remaining solids serve as the source of the bacteria-laden or "activated sludge," used in the aeration tank itself.

Trickling filters, likewise, make use of aerobic bacteria to decompose the organic constituents of primary effluents. In trickling filter processes, primary effluent is passed over rock beds that have been coated with a layer of biological slime. The slime contains aerobic bacteria that decompose the organic matter as the primary effluent passes over the rocks. The effluent from trickling filters is also passed to a settling tank to remove its residual solids.

Oxidizing ponds also use bacteria to process the primary effluent, which is passed into large ponds with surface areas of about one acre per thousand people served. The ponds are maintained in an aerobic state by mechanical aeration or through the photosynthetic activity of algae.

As a rule, excess sludge from secondary treatment processes are treated anaerobically, which converts most of it to methane and carbon dioxide. The residual, finally, may be used as land fill or incinerated for final disposal.

As can be seen in Table 4.4, typical liquid effluents from secondary

TABLE 4.4   Composition of a Typical Secondary Treatment Effluent[a]

Component	Concentration (mg/liter)
Gross organics	55
Biodegradable organics	25
Sodium	135
Potassium	15
Ammonium	20
Calcium	60
Magnesium	25
Chloride	130
Nitrate	15
Nitrite	1
Bicarbonate	300
Sulfate	100
Silica	50
Phosphate	25
Hardness (as calcium carbonate)	270
Alkalinity (as calcium carbonate)	250
Total dissolved solids	730

[a]From Horne 1978, page 695

treatment processes contain sizable quantities of potential water pollutants, and our subject elements are present in prominent concentrations. Within limits, passing the effluents to large bodies of water allows for final disposal by dilution. But more and more, it is becoming evident that tertiary treatment is necessary to minimize environmental damage, especially because of the large quantitities of plant nutrients the secondary effluents contain.

There are many approaches to tertiary treatment, detailed discussion of which is not appropriate to this book. Suffice to say here that the methods include, among others, lime coagulation, which removes residual suspended matter and phosphates, ion exchange and electrodialysis, which removes ionic constituents like carbonates, nitrates, and sulfates, charcoal adsorption, which removes adsorbable organics like aromatics as well as ionic constituents, and chemical precipitation with a number of different precipitants [Horne 1978, 693].

The last, incidentally, according to Horne, is capable of completely bypassing primary and secondary treatment. That is, it has been found that carefully controlled chemical precipitation is able to remove up to 90% of the suspended matter, 55% of the organic matter, and 90% of the bacteria in raw sewage. Chemical precipitation in combination with activated charcoal adsorption, thus, represents a potential complete sewage treatment process, eliminating need for biological processing.

5

Carbon, Nitrogen, and Sulfur in Air Pollution

In our earlier discussion of the history of environmental pollution, we introduced four major air pollution problems: 1) London smog—a consequence of burning high sulfur-bearing fuels—composed primarily of sulfur dioxide, its reaction products, and particulates; 2) photochemical smog— that mostly associated with discharge from internal combustion engines and composed primarily of oxides of nitrogen, carbon monoxide, and unburned hydrocarbons; 3) acid deposition—acidified rain, fog, and snow resulting from the interactions of oxides of nitrogen and sulfur with water vapor; and 4) ozone depletion—a consequence of upper atmosphere reactions between ozone and chlorofluorocarbons or oxides of nitrogen. In this section, we discuss the chemistry associated with each of them.

To appreciate the chemistry of air pollution, however, it is necessary to discuss, first, some elements of the meteorological phenomena that are involved in the dispersion of pollutants and, second, some elements of atmospheric chemistry. And we must introduce several new chemical entities, the free radical and singlet and triplet oxygen atoms; these are energetically excited chemical entities that are involved in nearly all the chemical phenomena underway in the atmosphere as well as being intimately involved in the formation of many air pollutants.

THE TEMPERATURE AND DENSITY OF THE ATMOSPHERE
AND AIR MOVEMENT

In contrast to the oceans of water on the surface of the earth, which are relatively uniform in temperature and density as a function of depth, the ocean of air above the surface exhibits a complex temperature profile and an exponential like decrease in density as a function of altitude. The temperature complexity derives from the influence of solar energy on the earth's atmosphere. The density variation arises because air, in contrast to water, is a compressible fluid, and its density is dependent upon the pressure exerted upon it as well as on its temperature.

The Atmospheric Temperature Profile

In our earlier discussion of the greenhouse effect, we alluded to the fact that the earth's atmosphere is essentially transparent to solar radiation. Were it not, we would not enjoy sunrises and sunsets. But at the earth's surface, substances exist and processes proceed that do absorb the sun's rays and, in so doing, generate heat. That heat is then re-emitted and is of appropriate wave length to be absorbed by atmospheric gases. As a result of the energy absorption–emission processes underway, the atmosphere assumes the altitude temperature profile that is illustrated in Figure 5.1.

As can be seen, the temperature of the lower atmosphere, the troposphere, is highest at the earth's surface and decreases with altitude.* In the upper atmosphere, however, the temperature profile inverts. That is, gas temperatures increase with altitude in the stratosphere owing to the ability of ozone to absorb solar ultra-violet radiation. Above the ozone layer, in the mesosphere, there is still another inversion, with temperatures decreasing because of the transparency of the gaseous components in residence there. And above that level, in the thermosphere, temperatures rise with altitude because the intensity and wave length of the solar ultraviolet radiation at these altitudes is such to be able to decompose the molecular oxygen and nitrogen present.[†]

The Atmospheric Density Profile

Air, like any gas, follows the well-known gas laws governing the relationships between pressure, volume, and temperature. That is, the product

---

*Of course, local fluctuations can occur, giving rise to anomalies in the generalized profiles we are discussing.

[†]It should be noted here that earthbound temperature is a measure of the energy transferred to a thermometer by impacting gas molecules. Thermodynamic temperature, which we are discussing here, is a measure of the kinetic energy of gas molecules. At the earth's surface, at atmospheric pressure, thermodynamic temperature and that which we observe with a thermometer are the same measures. At very, very high altitudes, where nearly vacuum conditions obtain, however, temperature is a statement of the velocity of gas molecules rather than the registration of a thermometer.

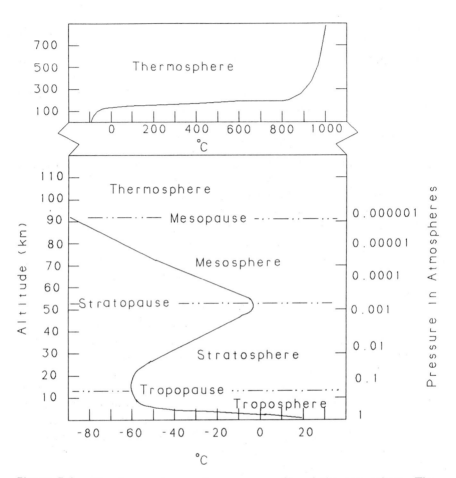

Figure 5.1   The temperature and pressure profile of the atmosphere. The transition zones between the troposphere, stratosphere, and mesosphere are called the tropopause, stratopause, and mesopause, respectively.

of its pressure and volume is proportional to its temperature and mass or

$$PV = nRT$$

where R is the gas constant and n is the quantity of gas in moles contained within the volume V. The product of n and the molecular weight of the gas is its mass, so

$$PV = \frac{mRT}{M_a}$$

where M_a is the molecular weight. Dividing by V and substituting the density, d, for m/V gives

$$d = \frac{PM_a}{RT}$$

showing the pressure and temperature dependence of gas density. Differentiating the last equation with respect to the altitude, z, one obtains

$$\frac{d(d)}{dz} = \frac{M_a}{R}\left[\frac{1}{T}\frac{dP}{dz} - \frac{Pdt}{T^2\,dz}\right]$$

relating the variation in air density with altitude to pressure–altitude and pressure–temperature profiles. But the air pressure at any altitude depends upon the weight of the air above it, i.e.,

$$P = zgd$$

where g is the gravity constanty. So, one can also write

$$\frac{dP}{dz} = -gd$$

Then, substituting for dP/dz, we can write

$$\frac{d(d)}{dz} = -\frac{M_a}{R}\left[\frac{gd}{T} + \frac{PdT}{T^2dz}\right]$$

If the temperature of the atmosphere were a constant, this last equation could be integrated to yield

$$d = d_o e^{-zgM_a/RT}$$

where d_o is the density of air at the earth's surface. That is, at constant temperature, air would exhibit a true exponential decline in density with altitude. The temperature profile of the atmosphere, of course, complicates the situation.

Inversions and the Movement of Air in the Atmosphere

Returning to the temperature dependence of gas density, we note from the gas law that the volume of a gas expanding at constant pressure is proportional to its temperature. That is

$$P_1V_1 = nRT_1$$

and

$$P_1V_2 = nRT_2$$

and, dividing one equation by the other, we get

$$\frac{V_1}{V_2} = \frac{T_1}{T_2}$$

Thus, if an increment of air is heated, its density will decrease; if it is heated to a temperature above its surroundings, buoyancy will force it in an upward direction. But as it travels up through decreasingly dense air, it will expand and, assuming no exchange of heat with its surroundings,* it will also cool. As long as the air increment is warmer than its surroundings, it will continue to rise. At some point, however, it will cool to a temperature equal to its surroundings and stabilize its motion. This process is, of course, the basis of convective mixing in the atmosphere, which one can think of as involving a very large number of air increments rising and falling as a function of the atmospheric temperature profile.

It is clear, therefore, that the local temperature profile of the atmosphere plays a key role in determining how pollutant bearing air increments mix and dilute. If the profile is such that temperature decreases with altitude faster than the temperature of the rising increment of air does, the increment will always be warmer than its surroundings and rise continuously, diluting and mixing its pollutants well. On the other hand, when the rising increment cools at a rate greater than the temperature profile of its surroundings, the increment stabilizes, allowing pollutants to concentrate in and below it. An inversion, of course, is an extreme example of just such a situation, where the rising increment meets a layer of warm air and either stabilizes or sinks to a point of equilibrium.

Global weather patterns, and the concurrent regional and global distribution of pollutants, are also caused by the behavior of rising and falling air increments. As is shown in Figure 5.2, warm air at the earth's surface, near the equator, tends to rise while the opposite is true at the poles. Thus, air masses at the equatorial region of the planet move to high elevations where they divide, part moving north and part moving south

---

*A valid assumption for a volume of air containing a much greater number of molecules than are able to exchange with its surroundings in the time frame of the expansion.

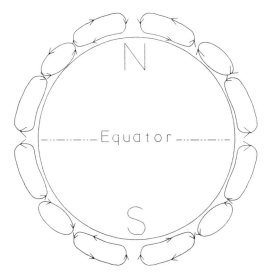

Figure 5.2   The north–south movement of air masses in the atmosphere.

toward the poles. At latitudes about 30° north and south, radiative cooling of the higher elevation masses causes them to fall. At the midlatitudes, between about 40° and 55°, in the temperate zones, influences of both equatorial and polar regions are in effect and lead to an upward motion of air masses at about 55° north and south latitudes. At higher elevations, those masses, like those rising from the equatorial regions, divide, with part moving back toward the equator and part toward the poles again [Seinfeld 1975, 128]. All the return paths, of course, are at the earth's surface, and all of the air mass movements are modified in direction, east and west, by frictional interactions with the earth's surface and by planetary rotation.

The motion of the air masses determines, in large measure, global climate and weather patterns. For example, latitudes near 30° north and south, where the air masses are falling, tend to be high pressure zones and warm, with generally pleasant weather patterns. This is because the falling air is compressively heated as it falls into regions of higher density air. Unfortunately, the compressive warming also gives rise to so-called subsidence inversions, which can contribute to air pollution problems. Subsidence inversion is one of the causes of the intransigence of the air pollution problem in Los Angeles, which is located at approximately 30°N latitude.

Inversions, of course, have other sources. On clear nights, for example, there can be significant cooling at the earth's surface because of heat radiating to the upper atmosphere and space.* The air near the surface, then, cools more rapidly than air at higher elevations, giving rise to an inversion that clears during the day when the sun warms the lower layer again. Cold fronts can cause inversions by sliding cool air beneath layers of warm air. Coastal regions can suffer daytime inversions because the air nearer the ground is warmed by the sun, rises, and is replaced by cooler air drawn from over the water by sea breezes. Inversions form over valleys and at the bases of hills as nighttime cooled air slides down, displacing warmer resident air at lower levels. Then, if a region is well shielded from breezes, as are valleys, the cold air pocket can not be readily cleared. The disaster at Donora, Pennsylvania, was one such example of an inversion over an industrialized valley community.

ATMOSPHERIC CHEMISTRY, EXCITED OXYGEN ATOMS, AND FREE RADICALS

Atmospheric Chemistry

Chemical reactions underway in the atmosphere are, in general, initiated by light. The atoms and molecules involved absorb photons of appropriate wave length, are exited to higher energy states, and, subsequently, dissipate their excess energy by participating in other reactions.
The primary photochemical reaction may be written

$$A + photon = A*$$

where the asterisk represents the excited species.

After photon absorption, the excited entity may follow a number of different paths to dissipate its energy. It may simply emit a photon of somewhat lower energy than the one that excited it, i.e., fluoresce

$$A* = A + photon$$

It may transfer its energy to some other entity in the form of heat

$$A* + M = A + M(hotter)$$

---

*It is not unusual for radiation cooling to freeze shallow pools of water at high elevations on cool clear nights, even when the ambient temperature is above freezing.

or it may engage in some kind of chemical reaction. Thus, if it is a molecule, it may dissociate into other molecular species or into highly reactive atomic species

$$A^* = A_1 + A_2 + A_3 + \ldots$$

or, if it is an atom, it may ionize

$$A^* = A^+ + \text{electron}$$

It may also react with another entity to form a new species, which might be one or more stable molecules or another excited entity

$$A^* + B = C + D + E + \ldots$$
or
$$A^* + B = A + B^*$$

The atmospheric gases all may engage in light induced reactions, each absorbing photons in its own unique spectral region. With regard to atmospheric chemistry, the important subsequent reaction of the excited product is generally dissociation into atoms. Molecular nitrogen, for example, absorbs only high energy photons of wave lengths less than 1200 Angstroms,* a region of the spectrum sometimes called the vacuum ultraviolet, and it dissociates into nitrogen atoms. In the thermosphere, therefore, molecular nitrogen serves as a filter to prevent that part of the spectrum from reaching the earth's surface. The resulting nitrogen atoms readily recombine, dissipating their reaction energy in the form of heat, which accounts for the very high thermodynamic temperatures of the thermosphere.

Molecular oxygen absorbs ultraviolet radiation ranging from the vacuum ultraviolet up to about 3000 Angstroms. Like nitrogen, therefore, it also serves to filter high energy radiation from the earth's surface. But its dissociation into atoms leads to reactions beyond recombination. The product atoms, themselves, can react with molecular oxygen to form ozone

$$O + O_2 = O_3$$

---

*An Angstrom is equal to 10^{-8} cm. The electromagnetic spectrum ranges from about 10^{-12} cm (cosmic rays) to 10^6 meters (radio waves). The most intense solar rays range in wave length from about 2000 Angstroms (the far ultraviolet) to about 18000 Angstroms (the far infrared).

It is the reaction of light activated oxygen in the upper atmosphere, with the subsequent formation of ozone, that is responsible for the stratospheric ozone layer.

Ozone, itself, absorbs radiation in two regions, between 2000 and 3200 Angstroms and between 4500 and 7000 Angstroms, the former band being the critical ultraviolet filter region of the stratospheric ozone layer while the latter band, being in the visual region of the spectrum, is involved in the formation of photochemical smog.

Excited Oxygen Atoms

When ozone absorbs radiation, it reverts to molecular oxygen and oxygen atoms, the latter of which may be in one of two electronically energetic states, depending upon the energy of the radiation that gave rise to the dissociation [Seinfeld 1975, 147]. Thus, the reactions

$$O_3 + photon(<3200 \text{ Angstoms}) = O_2 + O(^3P)$$

and

$$O_3 + photon(>3200 \text{ Angstroms}) = O_2 + O(^1D)$$

produce two reactive atomic species that may subsequently interact with other components of the atmosphere.

The two electonic states of the oxygen atoms may engage in different kinds of reactions with other components and are, therefore, generally distinguished from one another by consideration of their atomic spectra. Thus, the designation $O(^3P)$ refers to the triplet or ground state oxygen atom while the designation $O(^1D)$ refers to the singlet or electronically excited state.*

Free Radicals

Triplet oxygen atoms produced from the photodissociation of ozone are known to rapidly recombine with molecular oxygen to reform ozone in

---

*Recall that the normal state for oxygen is when its valence electrons occupy the p orbitals of the atom, thus the P designation for the ground state atom. But as has also been mentioned, an excited atom may have one or more of its valence electrons promoted to the higher energy d orbitals, thus the D designation for the excited atom. The superscripts designating the singlet and triplet states derive from the appearance of the absorption spectra of the two species; ground state oxygen atoms exhibit three closely spaced absorption lines while electronically excited oxygen atoms exhibit only a single line [Moelwyn–Hughes 1961, 247]

the troposphere. Singlet oxygens, however, react with water to form two units of the highly reactive hydroxyl free radical [Graedel 1980, 109]

$$H_2O + O(^1D) = 2HO\cdot \tag{1}$$

That is, when singlet oxygen reacts with water, one of the hydrogens splits away from the water molecule, carrying an electron with it, and combines with the excited oxygen atom (Figure 5.3a). Both of the resulting hydroxyl groups, then, remain with unpaired single bonding electrons that eagerly

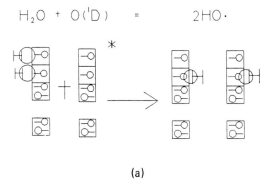

(a)

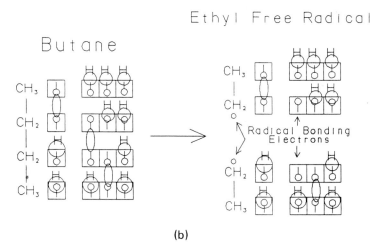

(b)

Figure 5.3   (a) The formation of hydroxy radicals from water and excited oxygen; (b) butane splitting to form two ethyl free radicals.

seek reaction with other species. Such odd electron containing groups are called free radicals and may be formed from other species in addition to hydroxyls.* Thus, in our earlier discussion of chemical bonding, we noted that covalent bonds were those in which valence electrons were shared by the bonding partners so as to satisfy inert gas configurations about each. Under proper conditions, however, covalently bonded molecules may be made to split similarly to water in its interaction with singlet oxygen, separating paired bonding electrons, and forming organic free radicals (Figure 5.3b).

In general, high energies are necessary to effect the formation of free radicals, energies associated with high temperatures and photochemical reactions. The atmosphere, therefore, is an excellent medium for radical formation because of the interaction of solar ultraviolet radiation with its components.

Radicals, incidentally, by virtue of their high reactivity, interact rapidly with other molecules, with possible reaction products including other radicals.

THE CHEMISTRY OF AIR POLLUTION

Sulfur Compounds

Atmospheric sulfur compounds include SO_2, SO_3, H_2SO_4, sulfate salts, and reduced compounds like H_2S and dimethlyl sulfide $(CH_3)_2S$. Recent estimates are that 30-50% of the sulfur resident in the atmosphere is anthropogenically generated as SO_2 while the remainder, derived from natural sources, is mostly generated in the form of reduced sulfur [Smil 1985, 399]. The large anthropogenic emissions of sulfur dioxide are supplemented in the atmosphere by the relatively rapid oxidation of the reduced sulfur compounds, making SO_2 one of the major sulfur species in the atmosphere. Local high concentrations of SO_2, of course, are the basis of so-called London smogs.

Earlier, in our discussion of the chemistry of sulfur, we noted that it and many of its compounds were generally oxidizable by molecular oxygen. However, reactions involving atmospheric sulfur species and molecular oxygen are too slow to account for their behavior in the environment. That is, the atmospheric residence times of sulfur species are

---

*Hydroxyl radicals should be distinguished from hydroxyl ions, the latter of which have inert gas electron configurations about their atoms.

relatively short, implying more rapid reaction than gas phase oxidations with molecular oxygen allows.

Investigations of the oxidation of sulfur species have indicated that there are several other oxidation modes possible. These include reactions with radicals, photochemical oxidation, and catalytic oxidation in liquid aerosols. So, for example, reduced sulfur is oxidized by ozone with the likely mechanism making use of water in the following reaction sequence [Graedel 1980, 113]

$$O_3 + light = O_2 + O(^1D)$$
$$O(^1D) + H_2O = 2HO\cdot$$
$$H_2S + HO\cdot = HS\cdot + H_2O$$
$$HS\cdot + O_2 = HO\cdot + SO\cdot$$
$$SO\cdot + O_2 = SO_2 + O(^1D)$$

Radical oxidation reactions with sulfur dioxide involve hydroxyls or hydroperoxyls, with the hydroperoxyl radical, $HO_2\cdot$, derived from the reaction of hydrogen atoms with water. There are several atmospheric sources of the hydrogen atoms. They include the reaction of atmospheric hydrogen* with triplet oxygen and with hydroxyl radicals

$$H_2 + O(^3P) = H\cdot + HO\cdot$$

and

$$H_2 + HO\cdot = H\cdot + H_2O$$

and the reaction of hydroxyl radicals with carbon monoxide

$$CO + HO\cdot = H\cdot + CO_2 \tag{2}$$

Whichever the source, the product hydrogen atoms then react with atmospheric oxygen to form the hydroperoxyl moiety

$$H\cdot + O_2 = HO_2\cdot \tag{3}$$

Sulfur dioxide may react with either radical species to form a bisulfite radical or sulfur trioxide

---

*The atmospheric concentration of molecular hydrogen is about 0.5 ppm.

$$SO_2 + HO\cdot = HSO_3\cdot$$

and

$$SO_2 + HO_2\cdot = SO_3 + HO\cdot$$

The SO_3 rapidly reacts with atmospheric water to form sulfuric acid and, while the exact mechanism is not known, the $HSO_3\cdot$ is thought to also play a role in the formation of acidified water [Graedel 1980, 114].

Photochemical oxidation of SO_2 produces excited triplet and singlet SO_2 molecules

$$SO_2 + photon(2900–3400 \text{ Angstroms}) = {}^1SO_2$$

and

$$SO_2 + photon(3400–4000 \text{ Angstroms}) = {}^3SO_2$$

which enter into reaction with other atmospheric components to form SO_3. In particular, the triplet SO_2 is thought to react with atmospheric oxygen to form SO_3 and atomic oxygen.

Interestingly, when hydrocarbons and oxides of nitrogen are introduced into mixtures of air and sulfur dioxide, the photochemical oxidations are found to accelerate [Seinfeld 1975, 192], highlighting the synergistic influences of other pollutants in the formation of atmospheric sulfuric acid.

Catalytic oxidation of sulfur dioxide occurs when it is dissolved in water containing metal salts such as the sulfates and chlorides of manganese and iron. Observed oxidation rates have been 10 to 100 times those of the oxidation of SO_2 in clean air. Since iron and manganese salts frequently exist as suspended particulates in air, they may serve, under conditions of high humidity, as condensation nuclei for water droplets. Ambient SO_2 dissolving in those droplets is then rapidly oxidized directly to sulfuric acid.

Nitrogen Compounds

Oxides of nitrogen are intimately involved in the formation of photochemical smog. They are a contributing factor in acid rain; and they may contribute, along with chlorfluorocarbons, to the depletion of the ozone layer.

The earth's atmosphere is about 80% molecular nitrogen and contains about two percent of the planet's total nitrogen. The second most abundant of the atmospheric nitrogen compounds is nitrous oxide, which is uniformly mixed in the troposphere and present to the extent of about 0.33 ppm. In the stratosphere, its concentration is about 0.1 ppm [Graedel 1980, 108]. However, the low concentrations belie the importance of N_2O as a component of the atmosphere, especially in the stratosphere.

Recall that molecular oxygen is photodecomposed by ultraviolet rays of less than 3000 Angstroms to singlet oxygen atoms; these react with N_2O to form nitric oxide

$$N_2O + O(^1D) = 2NO$$

This reaction is a major N_2O disappearance route in the atmosphere [Hauck 1984, 106]. But because little 3000 Angstrom radiation penetrates lower elevations, the reaction does not occur in the troposphere. As a result, nitrous oxide behaves like an inert gas with a stable concentration in the troposphere. Virtually all of its consumption occurs in the stratosphere where the gas takes an active role in the nitrogen cycle. There, it initiates a natural catalytic reaction sequence that helps to limit the stratospheric concentration of ozone. That is, the nitric oxide produced by photodecompostion of N_2O engages in the following cyclic reaction sequence

$$NO + O_3 = NO_2 + O_2$$

and

$$NO_2 + O(^1D) = NO + O_2$$

and, thus, destroys stratospheric ozone. Competitive reactions between the nitrogen dioxide produced in the cycle and hydroxyl radicals, produced from water and singlet oxygens, control the extent to which stratospheric ozone is naturally depleted by N_2O. That is, the reaction

$$NO_2 + HO \cdot = HNO_3 \qquad\qquad (4)$$

removes NO_2 from the cycle.

But as can be seen, excess concentrations of N_2O, generated by increased fertilizer use, for example, can ultimately diffuse to the stratosphere and contribute to the depletion of the ozone layer by serving as a source of excess NO.

The most important reactions of nitrogen in the troposphere, especially as related to air pollution, involve the formation of their oxides from combustion. Nitric oxide is produced in high temperature combustion processes like those associated with internal combustion engines and steam generated electric power. The nitric oxide then serves as the precursor to nitrogen dioxide. In concert with peroxyhydroxyl radical produced through the reactions numbered 1, 2, and 3 above, it is oxidized to NO_2*

$$NO + HO_2\cdot = NO_2 + HO\cdot \qquad (5)$$

Nitrogen dioxide then, in addition to being one cause of the brown color of smog, serves as a photochemical trigger for the formation of other pollutants. That is, the photodissociation of NO_2 results in the production of ozone according to

$$NO_2 + photon = NO + O$$

and

$$O + O_2 = O_3$$

and the ozone reacts with unburned hydrocarbons, as is discussed below, to form many of the noxious products associated with smog.

Nitrogen dioxide may also react with hydroxyl radicals (produced, for example, according to Equations 1 and 5) to form nitric acid (Equation 4); and nitric acid, in additition to being a pollutant in its own right, contributes to the formation of acid rain.

Carbon Compounds

Carbon monoxide is, of course, a major product of combustion processes, including those involved in internal combustion engines, power generation, and incineration. Its role as a pollutant in its own right, because of its high toxicity, needs no further discussion; and we have already mentioned the part it plays in the production of hydrogen atoms and the subsequent formation of peroxyhydroxyl radicals (Equations 2 and 3). But carbon compounds beyond carbon monoxide are also major contributors to air pollution. Unburned hydrocarbons from auto exhausts, industrial solvents, effluents from chemical manufacturing plants, etc., all contribute by interacting with the products of other photoinduced reactions.

The three major reaction mechanisms that atmospheric carbon compounds engage in involve singlet oxygen, hydroxyl radicals, and ozone.

---

*There is always a small amount of NO_2 associated with the NO generated in combustion processes. The major source of NO_2 in air pollution, however, is from postcombustion oxidation of NO by peroxyhydroxyl radicals and similar moieties.

All of these react with organic molecules to form organic radicals. These radicals are precursors to, among other things, carbon-containing oxidants which, like ozone and nitric acid, are the air pollutants that inflict large scale injury to plants and materials. They are also the obnoxious carbon-containing products that lead to the eye and throat irritating qualities of photochemical smogs.

Singlet oxygen, produced from the photodissociation of molecular oxygen, reacts with hydrocarbons to produce organic radicals as follows

$$RH + O(^1D) = R\cdot + HO\cdot$$

And the resulting hydroxyl radicals, as well as those produced from the reaction of water and singlet oxygen, also react with hydrocarbons

$$HO\cdot + RH = R\cdot + H_2O$$

to produce organic radicals.

The organic radicals then react rapidly with oxygen to form peroxy radicals

$$R\cdot + O_2 = R - O - O\cdot$$

which, in turn, react with NO_2 to form alkylperoxynitrates

$$R - O - O\cdot + NO_2 = R - O - ONO_2$$

one class of oxidizing air pollutants.

The reaction mechanism for ozone's formation of radicals is more complicated. Ozone does not extract hydrogens from organics. More typically, it forms radicals by reacting with unsaturated hydrocarbons like ethylene or propylene, forming oxygen addition compounds according to

$$H_3C = CH_3 + O_3 = \begin{array}{c} H_3C - CH_3 \\ | \quad\quad | \\ O \quad\quad O \\ \diagdown\;\diagup \\ O \end{array}$$

which then decompose to form aldehydes and unstable diradicals of the form

$$H-C\overset{\overset{\displaystyle H}{\diagup}}{\underset{\diagdown O}{}} \quad \text{and} \quad H_3C-\overset{\overset{\displaystyle H}{|}}{\underset{\bullet}{C}}-O-O\bullet$$

The diradicals, in turn, decompose to other radicals, examples of which are

$$HO\bullet \quad \text{and} \quad H_3C-\underset{\bullet}{C}O$$

These and other product radicals react with other organic molecules in countless ways to form countless other kinds of carbon pollutants.

The aldehydes formed in the ozone reactions also contribute to air pollution by serving as the precursors to another class of organic oxidants called peroxyacyl nitrates. An example of one is peroxyacetyl nitrate (PAN), the well known lachrymator, or eye irritant, found in photochemical smog.

PAN is formed as a consequence of the reaction of the aldehyde with hydroxyl radical

$$H-C\overset{\diagup O}{\underset{\diagdown H}{}} + HO\bullet = H-\underset{\bullet}{C}\overset{\diagup O}{} + H_2O$$

followed by reaction with oxygen

$$H-\underset{\bullet}{C}\overset{\diagup O}{} O_2 + \quad H-C\overset{\diagup O}{}-O-O\bullet$$

to form the peroxyacyl radical. And the peroxyacyl radical, in turn, reacts with NO_2 to form PAN

$$H-C-O-O\bullet + NO_2 = H-C\overset{\diagup O}{}-O-O-NO_2$$

There are a multitude of other photochemically induced reactions that have been recognized as contributing to air pollution and the formation of smog, discussion of which is beyond the scope of this book. The reactions presented in the foregoing discussion, however, do represent a sampling of those most intensely studied and are thought to be among the most significant.

Ozone Depletion and Chlorofluorocarbons*

Chlorofluorocarbons (CFCs) are a class of synthetic aliphatic carbon compounds that have had their hydrogens totally substituted with chlorine or fluorine. They are popularly known by the trade name Freon and are ubiquitously used as aerosol propellants, refrigerants, and blow-molding and foaming agents.

Large scale production of CFCs began in the 1930s, and their release into the atmosphere increased progressively until 1977 when the EPA placed regulatory limits on their use. Worldwide annual releases then fell from about 750 million pounds per year in 1976 to about 600 million pounds per year in 1983. Worldwide cumulative releases between the time CFCs were introduced and 1983 have been in excess of 12 billion pounds [Environmental Quality 1984, 713].

CFCs are, for all practical purposes, chemically inert, but they are subject to photodecomposition by ultraviolet radiation in the 2250 Angstrom region, i.e., the spectral region that penetrates to the stratosphere, but not the troposphere. As such, they have extremely long lifetimes in the troposphere—of the order of 30 to 100 years [Graedel 1980, 127]—and are distributed fairly uniformly there, but decline in concentration with altitude in the stratosphere due to their photodecomposition.

In 1974, it was suggested that CFCs represented a threat to the ozone layer [Molina and Rowland 1974]. The proposed mechanism for ozone depletion was similar to that of nitrous oxide. That is, the enormous releases of CFCs would ultimately drive them to diffuse into the stratosphere in high concentrations where their reaction products could attack the ozone. Typical CFCs like fluorotrichloromethane and difluorodichloromethane photodecompose and produce chloroflourocarbon radicals and chlorine atoms

$$CFCl_3 + photon = CFCl_2 \cdot + Cl \cdot$$

and

$$CF_2Cl_2 + photon = CF_2Cl \cdot + Cl \cdot$$

The chlorofluorocarbon radicals recombine to form new compounds. The chlorine atoms, however, can initiate a catalytic reaction chain that can

*Russow [1980] has presented a good general discussion of chlorofluorocarbons.

lead to ozone depletion

$$Cl\cdot + O_3 = ClO\cdot + O_2$$

and   (6)

$$ClO\cdot + O = Cl\cdot + O_2$$

However, until the recent observation that Antarctic ozone was depleting to a point of disappearance for several months at a time, there was great debate about the influence of these reactions on the ozone layer [Zurer 1987]. The counter argument was that 1) the chlorine monoxide produced in Equation 6 would be promptly removed from the cycle by stratospheric nitrogen dioxide (derived from N_2O) with the production of chlorine nitrate

$$ClO\cdot + NO_2 = ClONO_2$$ (7)

a compound that does not attack ozone and 2) that chlorine atoms would also be quickly removed from the cycle by reaction with methane, another gas that is essentially inert in the troposphere but can diffuse to the stratosphere to react with chlorine atoms to produce hydrochloric acid

$$Cl\cdot + CH_4 = HCl + CH_3\cdot$$ (8)

The debate was finally resolved with the recent observation that chlorine nitrate reacted with HCl in the stratosphere heterogeneously, i.e., at ice particle surfaces, to form molecular chlorine and nitric acid [Zurer 1987], and that the nitric acid produced was then absorbed by the ice

$$ClONO_2 + HCl = Cl_2 + HNO_3$$

and

$$HNO_3 + ice = dissolved\ nitric\ acid$$

while the chlorine remained in the gas phase where it was easily photolyzed to chlorine atoms.

The depletion sequence, therefore, is thought to be initiated with the production of chlorine atoms from the photodissociation of the CFCs followed by formation of chlorine nitrate and the reaction sequence

$$ClONO_2 + HCl = Cl_2 + HNO_3$$
$$Cl_2 + photon = 2Cl\cdot$$
$$Cl\cdot + O_3 = ClO\cdot + O_2$$
$$ClO\cdot + NO_2 = ClONO_2$$

with the net reaction being

$$HCl + 2O_3 + NO_2 = HNO_3 + 2O_2 + ClO\cdot$$

and always moving to the right because of the dissolution of the nitric acid by ice crystals.

Part III

The Analysis of Carbon, Nitrogen, and Sulfur in the Environment

Any discussion of the analysis of environmental pollutants would be less than complete without mention of criteria that are used for selection of analytical methods. Clearly, in analytical chemistry, as in other endeavors, there is almost always more than one way to skin the cat. As we shall see, therefore, there are preferred analytical approaches in environmental analysis. Before an analysis may be conducted, however, consideration must be given to the nature of the sample being investigated. Quantitative analysis requires that samples being analyzed be truly representative of the space from which they are taken. Since the environment is uniquely non-uniform and dynamic, compared with, for example, the product of a manufacturing process, environmental analysis demands particular care and planning in sampling. Tied to sampling, also, is the extent to which a given analytical method provides reproducible results. The reproducibility of an analysis must be better than the expected variance in sample character in order to distinguish differences among samples. In the first chapter of this section, therefore, discussion has been directed to method selection, sampling procedures, sampling statistics, and error analysis.

With respect to the remaining chapters of the section, which are discussions of analytical methods, organization has necessarily been arbitrary. Since this is a monograph on carbon, nitrogen, and sulfur pollutants, a natural division might have been a discussion of methods appropriate to each of the subject elements, individually. However, many of the instruments, detectors, and separation procedures that are incorporated into the applicable analytical technology are common to all three elements. So, our discussion of methods is organized along the lines of principles instead, identifying each of the methods with individual pollutant determinations as they are applicable. The methods discussion is, therefore, divided into five chapters devoted to physical methods, electrochemical methods, chemical methods, chromatography, and bioanalytical methods.

Physical methods make use of one or more of the physical properties of the analytes for separation and/or quantitation, electrochemical methods are based on measurement of characteristic voltages or current flows that are associated with chemical transformations, and chemical methods make use of chemical transformations as the primary basis of separation and quantitation. As will become obvious, however, the lines between chemical and physical methods are frequently obscured, because methods that are fundamentally physical are often employed as detection and measurement procedures for chemically modified analytes.

Chromatography is treated independently because it is, first and foremost, a separation technique; it becomes a powerful analytical tool by virtue of its use of a number of different physical and even chemical methods for detection and quantitation. As will be seen, some of these are unique to chromatography; others are applications of more general analytical methods covered in the other sections of the discussion.

Finally, the chapter on bioanalytical techniques has been included despite the fact that, as of the date of this writing, no bio-methods had yet been sanctioned by any regulatory agencies nor had any become commercially available. The chapter has been included because, as will be seen, bio-methods, particularly enzymatic and immunological methods, are exquisitely specific and sensitive, require relatively simple manipulations, and are eminently applicable to analysis of a number of environmental pollutants. It is this writer's opinion that their general application to environmental monitoring is inevitable.

6

The Use of Analytical Methods

SELECTION OF ANALYTICAL METHODS

It has been said that there are about 50 broad classifications of methods that are used in analytical chemistry. Within those classes are hundreds of procedures that can be applied to pollutant analysis. In establishing a new method, the experienced analytical chemist will review the broad classes, searching for types of analyses he judges to be adequately specific* and sensitive[†] for his needs. He will then select specific procedures within those types that he can adapt to the equipment and facilities unique to his laboratory. To the analytically uninitiated, however, selection of methodology for quantitating compounds of carbon, nitrogen, and sulfur from the wealth and diversity of methods within those 50 broad classifications can be overwhelming.

Fortunately, at least for environmental monitoring in the United States, the range of possible methodologies can be narrowed, because reference methods for the analysis of the criteria air pollutants[‡] and for sulfates, nitrates, and toxic organics in water have been defined by the

---

*Analytical specificity is the extent to which the method directs itself to the analyte of interest, isolating it from possible interferrants. We shall discuss, later, analytical methods that are fundamentally separation techniques combined with generalized detection methods.
[†]Analytical sensitivity is lowest level of analyte that may be expected to be quantitatively determined.
[‡]Oxides of sulfur and nitrogen, carbon monoxide, hydrocarbons, ozone, particulate matter, and lead.

125

EPA. These reference methods may themselves be used as monitoring procedures or they may serve as the criteria by which other methods are judged to be equivalent in efficacy and acceptable to the EPA as substitute procedures. Although obtaining regulatory sanction for an equivalent method is eminently possible, it is also a somewhat drawn-out and complicated procedure. Therefore, the number of methods available to the analyst wishing to conform to regulatory requirements is limited.

Thus, while the experienced analyst may be able to adapt or develop a method that may ultimately be sanctioned as equivalent, the uninitiated, in selecting his method, must give priority consideration to whether or not it is an approved equivalent. This is true, incidentally, regardless of one's reporting obligations to regulatory agencies, because the fact of equivalence establishes the method as having adequate specificity and sensitivity for its application; it can, therefore, be confidently applied internally. For the uninitiated, therefore, the primary literature reference is the Code of Federal Regulations, Title 40 (40 CFR). Parts 50 and 53 of 40 CFR are directed to air pollutants. Part 50 defines primary reference methods for the criteria air pollutants and Part 53 defines procedures that must be followed to win approval of new reference or equivalent methods. Part 136, dealing with water pollution, lists reference and equivalent methods for inorganic and organic pollutants along with the procedure to be followed to win appproval of alternatives at national, regional, or local levels.*

Although the Code is revised annually, it does not necessarily include all the currently approved methods, nor does it list equivalent methods that have won approval at the regional EPA level. However, EPA regularly publishes revised lists of designated reference and equivalent methods for measuring air and water pollutants, so it is worthwhile inquiring at the local or regional EPA office for the most current method recommendations. The details of equivalent methods are published in the Federal Register and in the Analytical Quality Control Newsletter of the EPA's Monitoring and Support Laboratory in Cincinnati, Ohio.

An additional and excellent reference, specifically directed to air pollution, is the Quality Assurance Handbook for Air Pollution Measurement Systems⁺ published by the Environmental Monitoring Systems Laboratory

---

*The U.S. is divided into a number of EPA regions, each headed by a Regional Director. Federal regulations permit Regional Directors to approve equivalent methods for use within their own regions.
⁺These handbooks may be obtained at no charge from the U.S. Environmental Protection Agency, Center for Environmental Research Information, 26 W. St. Clair Street, Cincinnati, OH 45268.

of the EPA at Research Triangle Park, N.C. This is a massive work in five volumes that covers all aspects of EPA recommended and designated procedures, extending from sampling protocols to analytical methodologies for ambient air, stationary sources, and acid rain.

Manufacturers of commercial analytical devices are also rich sources of information relative to methods and regulatory requirements. Caution must be used, however, in relying on manufacturers' information, since it is clear that their primary motivation is to sell product.

Other considerations that should be given to selection of methodology include budget, level of personnel sophistication, and frequency of analysis. Starting with the last, very labor intensive methods can be tolerable, and may even be more cost effective than sophisticated instrumental methods if analysis requests are relatively infrequent. On the other hand, some of the more labor intensive methods, as many of EPA's reference methods are, may also require application of relatively complex laboratory manipulations. Thus, the level of technician sophisitication available may dictate acquisition of the more expensive automated methods. Budget, of course, is the ultimate determinant of the method selected. But a good rule of thumb, unless one is in the business of researching and developing methodology, is to acquire the least sophisticated, least expensive apparatus available, in keeping with one's analytical abilities and needs.

In choosing a supplier for a given method, one should consider what he provides in service and technical support as well as in apparatus. Equipment suppliers vary from very small to very large, from manufacturers with their own sales forces to those who use manufacturer's representatives* or laboratory supply houses for distribution. Each supplier will highlight the distinctions between his and his competitors' devices, but careful examination will usually reveal that these are more in degree than in kind. That is, a sulfur analyzer, based on an EPA approved method, will determine sulfur, regardless of whom manufactures it. And if it is an approved method, analysis parameters like specificity and sensitivity will also be adequate. Distinctions, if any, will center around convenience features rather than perfomance, and price will usually pace the number of convenience features provided.

In this writer's opinion, apart from service, which one should be able to take for granted, the most important feature any supplier can provide

---

*A manufacturer's representative is an independent businessperson who may represent several different equipment manufacturers. His income is derived from commissions he receives on sales of equipment and its associated supplies. The reputable representatives are frequently technically sophisticiated and will try to place equipment properly. The less-than-reputable ones will try to place the highest price devices.

to the nonanalytical chemist is technical support. And despite the possible protestations of sales people, who sometimes think they can provide adequate support, access to the manufacturer's technical staff is a critical component of support. One of the best measures of the quality of the support that may be provided is the determination of whether the supplier maintains an on-going applications research activity, and whether the customer can have direct access to it. In that regard, though one might pay somewhat higher prices for the equipment, the larger manufacturers may be the best suppliers for the less sophisticated analysts.

Possibly the worst source of equipment is a large laboratory supply house. These organizations generally serve to simply warehouse and distribute other manufacturers' products. They rarely have any internal support capability and will direct technical questions and problems back to the manufacturer, thus serving as an intermediary block in trying to get problems resolved or technical tidbits transmitted. For similar reasons, one should also be wary of importers who themselves do not maintain laboratory and service facilities, despite the good reputations that the manufacturers they represent may have.

In order of support expectations from suppliers, this writer's experience places the larger equipment manufacturers with internal applications research facilities and sales staff first, smaller domestic manufacturers using manufacturer's representatives as sales staff next, and importers and laboratory supply houses last.

SAMPLING FOR CHEMICAL ANALYSIS

It is an *a priori* truth that an analytical result is no better than the extent to which the analyzed sample is representative of the material whose composition is being determined. Thus, the procedure used to collect samples is as critical as the analytical methodology itself, and sampling the environment for purposes of pollution analysis is particularly complicated.

In contrast to the product of a chemical factory, for example, where bulk material is well confined and may be manipulated to achieve a degree of homogeneity, the environment is intrinsically heterogeneous, both spatially and temporally. As a consequence, the sampling procedures used to achieve rational representation of pollution status are, more often than not, unique to the specific universe being analyzed. As examples, quiescent waters may require sampling at multiple depths at different surface locations, and always with awareness of the pollutant concentrating effects of the air–water interface (see Chapter 4, p. 88). Rivers may require similar spatial sampling, but as a function of time as well, to account for contaminants that may intermittantly flow from tributory streams. Air samples around factory stacks must be taken in time and space patterns that reflect the strength and direction of prevailing winds, the topography, and the amount of incident solar radiation, as well as manufacturing activity, which may cause sporadic rather than continuous discharge of contaminants. And drinking water or sewage effluents exiting processing plants must be sampled as a function of time and as a function of the cross section of the effluent pipes, because flow characteristics in pipes can produce contaminant concentration gradients.

Additionally, the rate at which pollutants enter and disperse throughout the sampling space may dictate sampling procedures. That is, if sampling time is long with respect to the generation and dispersion of a pollutant, a single sample may be a reasonable representation of the bulk. Contrariwise, if the sampling time is short, it may only represent a fragment of the entering contaminant, and time averaging of multiple measurements may be required to obtain an adequate representation.

How one defines specific sampling protocols, then, may be a problem in limnology for lake or river analysis, in micrometeorology for air analysis, or in chemical engineering for effluent analysis. That is to say, sampling protocols are not the province of the analyst alone. One must know something about the gross behavior of the bodies being analyzed to define sampling patterns. What the analyst must specify, however, is the nature of the container used to take samples, so as to avoid sample-container interaction, the amount of sample required in keeping with the sensitivity of the method, and the time constraints between sampling and analysis so as to minimize sample instability losses.

Although the EPA makes specific demands relative to some of these factors, the American Society For Testing and Materials (ASTM) can serve as an excellent supplementary source of information about sample handling. That is, the EPA lists specific container materials, acceptable preservation methods, and maximum sample holding times for analyzing our subject elements in water [40 CFR Pt. 136.3 1985, 258; Shelley 1977]; and it spells out detailed sample collection protocols that are used in concert with its reference methods for analyzing ambient air pollutants [40 CFR Pt. 50 1986, 528; Quality Assurance Handbook 1976,1977]. The ASTM, on the other hand, provides a number of different procedural recommendations for sampling water [ASTM 1987, V 11.01, 11.02] as well as the atmosphere [ASTM 1987, V 11.03]. Standard Practices for Sampling Water (D 3370-82), for example, is a discussion of sample collection methods that includes grab sampling, composite sampling, and continuous monitoring. Standard D 1192 is directed to collection apparatus, and Standard D 4515-85 is a discussion of ways to estimate maximum holding times for water samples designated for organic analysis. The appropriate standards for air include D 1357-82 and D 3249-79, which are directed to ambient atmosphere sampling, and D 1605-60, which is concerned with general sampling for analysis of gases and vapors and includes discussions of collection-canister types and methods to concentrate samples by adsorbing them on high surface solids or by condensation.

STATISTICS OF SAMPLING

The primary purpose of a sampling protocol is to provide representative samples of the space being investigated. Properly designed, it adds validity to the overall analysis of the data and to the conclusions drawn from it. The sampling protocol itself, however, does not eliminate errors associated with collection procedures and analytical determinations. That is, all analytical procedures are subject to errors arising from intrinsic irreproducibility in measurement as well as in sample handling itself. If these errors are large relative to differences that are being sought among the samples being analyzed, then conclusions drawn from the data relating to the differences may be suspect.

As an example, consider analyzing a lake fed by undergound springs that is being contaminated by nitrates, with the source of the contamination being a subject of dispute. One would like to know first, what the overall nitrate concentration is so as to be able to assess the threat to the lake's health and second, where the major contaminant entry points might be. One plan to achieve both objectives might be to construct a grid over the surface of the lake and sample at multiple depths in selected areas of the grid. Then one could average the data to obtain the overall nitrate concentration; and the sampled grid would reveal any patterns of entry into the lake by displaying clusters of higher-than-average nitrate concentrations near the entry points. Conclusions about the lake, of course, rest on the assumptions that the sample average will be a proper representation of the true average concentration and that the concentration differences between the clusters and the bulk of the lake will be larger than the intrinsic sampling and measurement errors. Unfortunately, in most real situations, the inherent heterogeneity of the sample space can obscure these assumptions. Statistical analysis provides a mechanism to test their validity.

The mechanism is derived from sampling theory, which itself is based on the assumption that measurements of a property of a *uniform* sampling space will always tend to cluster about an average value, with measurements that differ only a small amount from the average appearing more often than those that differ by larger amounts. If one were able to make an infinite number of measurements and plot their values as a function of freqency, one would obtain a so-called normal distribution of values, i.e., the bell-shaped curve in Figure 6.1. The maximum in the normal distribution would be the true value of the measured property because it would represent the average of all possible measurements, and the width of the distribution curve would represent the variability in the measurements. Thus, the normal distribution represents a theoretical ideal for an infinitely large set of measurements burdened with totally random fluctuations in values.

The normal distribution is described by the equation

$$y = \frac{1}{\sigma 2\pi} \exp\left[\frac{-(x - x_{av})^2}{2\sigma^2}\right]$$

where y is the frequency of appearance of any measured value, x is the measured value, and x_{av} the average or true value for an infinitely large sample.

A measure of the quality of the average of a normal distribution of a sample population is the width of the distribution curve. One way to

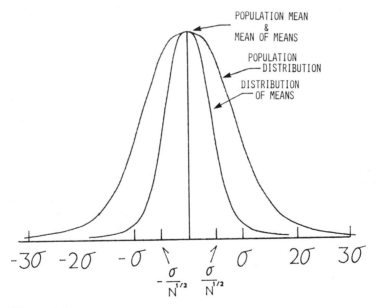

Figure 6.1   The normal distribution of a population and the distribution of the means of samples selected from it. Note that the standard deviation of the means is $\sigma/N^{1/2}$.

express the quality is in terms of the average of the sum of the absolute values of the deviations from the mean of the distribution*

$$Q = \frac{\Sigma_i |(x_{av} - x_i)|}{N}$$

But more useful measures, provided by mathematical statistics, are the variance, which is defined as

$$V = \frac{\Sigma_i (x_{av} - x_i)^2}{N}$$

and σ, the standard deviation, which is given by the square root of the variance

$$\sigma = \left[\frac{\Sigma_i (x_i - x_{av})^2}{N} \right]^{1/2}$$

The value of σ has an interesting property which is, in effect, the basis of the statistical analysis of normal populations. That is, it can be shown that 68.27% of all the measured values of a normal distribution fall within one standard deviation (1σ) of the average value, that 95.45% of them fall within two standard deviations (2σ), and that 99.73% fall within three standard deviations (3σ) of it. Thus, in a truly normal distribution, one can say that there is approximately a 50% chance (68:32 or about 2:1) that any measured value will be within one standard deviation of the mean, a 95% chance (19:1) that it will be within two standard deviations, and virtual certainty (about 350:1) that it will fall at least within three standard deviations.

Real life measurements, of course, encompass only finite sample sets. But these same probability arguments may be applied to them as well. Each of the finite sets can be considered a sample of the infinite population of measurements that we've been discussing, and each of the sample sets will have its own average and standard deviation, which may be examined in the context of the infinite set. That is, it can be shown that the distribution of the means of a large number of real sample sets of, say, ten measurements each, is also normal, with their values falling on a curve

---

*The absolute values of the deviations are used because the algebraic sum of the deviations from the average of a normal distribution would be zero.

whose maximum (the mean of means) is close to or equal to the mean of the infinite set of measurements. But the standard deviation of the means of the sample sets is smaller than that of the individual measurements. In fact, it can be shown that the standard deviation of the means is given by

$$\sigma_{av} = \frac{\sigma}{N^{1/2}}$$

where N is the number of measurements within each sample set (Figure 6.1). So, if one were to take a single sample of 10 measurements, of our contaminated lake for example, one could conclude that about 50% of time, its mean would lie somewhere within about $\sigma/3$ of the true value, and that it was fairly certain (about 350:1) that it would always lie within 1σ of the true value, so long as the fluctuations among its individual measurements were truly random. But we still have no measure of the degree to which the mean of the single sample may compare to the true value because we do not know the magnitude of σ. However, an estimate of its value is accessible if one considers, again, the large set of samples of ten. Each of these sample sets exhibits its own standard deviation, which we can denote by s and which is given by

$$s = \left[\frac{\Sigma_i (x_i - x_{av})^2}{N} \right]^{1/2}$$

And it can be shown that the values of s, plotted as a function of frequency, also yield a normal distribution [Baird 1964, 34], one that is centered about the value of σ for the infinite set and is given by

$$\sigma_s = \frac{\sigma}{[2(N - 1)]^{1/2}}$$

Then, reasoning as we did for the mean of our single sample of ten, we can conclude that there is about a 50% chance that any single value of s will lie within a range of $\pm\sigma/4$, or about $\pm0.25\sigma$ of the standard deviation of the infinite set of measurements, and that it almost certainly will lie within a range of $\pm3\sigma/4$, or about 0.75σ. Clearly, the larger the number of measurements in the sample set, the narrower will be the range in which s may fall relative to σ (for $N = \infty$, of course, $s = \sigma$). Table 6.1 illustrates the ranges of s that may be expected at the 95% confidence level ($2\delta_s$) as a function of N. Thus, for a large sample, it is not unreasonable to

TABLE 6.1   Expected Spread in
Values of s as a Function of Sample
Number

N	$\dfrac{2}{[2(N-1)]^{1/2}}\sigma$
2	1.4
3	1
4	0.83
5	0.71
6	0.62
7	0.59
8	0.56
9	0.50
10	0.48
15	0.38
20	0.31
50	0.20
100	0.14

use the value of s as an estimate of the value of σ. In fact, statisticians call the relationship

$$s = \left[\frac{\Sigma_i(x_i - x_{av})^2}{N-1}\right]^{1/2} = \text{Estimate of } \sigma$$

the best unbiased estimate of the value of σ for finite samples of any size [Baird 1964, 35].*

We may now sample our contaminated lake at, for example, ten grid locations and five depths and compute a mean concentration value which we can, at the 95% confidence level (Table 6.1), anticipate to be no further away from the true value than about ± 0.2 of the standard deviation of our sample of 50; and we can scrutinize our individual results in terms of the average value and its standard deviation, using the latter as a test for the likelihood than any individual result is either a random fluctuation or some extraneous effect. That is, we can compare any individual measurement with the mean and argue, if its value falls outside the 2σ region,

---

*Note, that $(N-1)$ instead of N is used in the denominator, which we state without proof. However, this is an especially important addition if N is small, because it biases the estimate of σ to a slightly higher value to reflect expected differences between the true mean and the averages of small samples.

that there is only a 5% chance that it is a chance fluctuation, that it is, more likely, different from the mean; and we may use that conclusion to search for nitrate entry patterns.

There are other statistical tests that may also be applied to our lake data, details of which may be found in several excellent introductory texts on statistical analyis of measurements [Youden 1951; Baird 1962; Rabinowicz 1970]. They include the Student T, which is a test of the equality of averages. Thus, for our contaminated lake, we might divide the grid into large sections, calculate means for each, and use the Student T to determine the likelihood of their being the same or different from one another. The chi^2 test is a comparison of an observed sample data distribution with a theoretical normal distribution. It yields the likelihood that the observed distribution is truly random. Thus, chi^2 values for the distribution of nitrate in our contaminated lake might suggest a non-normal data distribution and the presence of entry patterns.

Regardless of which tests may be applied to the data, however, it is important to remember that statistical tests, in fact statistical analysis generally, is based on a primary assumption that data spread is a consequence of random events. Thus, the best one can ever do in statistically analyzing real measurements, in terms of trying to assess their validity, is to compare them to theoretical distributions and ask whether the observed variability is more likely to be the result of random fluctuations than it is to be caused by some systematic bias or effect. This question is a statement of the statistician's *null hypothesis* which says, in effect, that variability in data may be attributed to either random fluctuations or systematic bias, but not both. Tests of statistical significance address the null hypothesis only, i.e., provide a probabilistic yes/no answer as to whether observed differences in data are the result of chance or not. They cannot differentiate degrees of randomness or systematic biases.

ANALYSIS ERROR

Analytical methods carry their own intrinsic errors, which contribute to the data spread one may expect sampling a normal population. However, one can estimate the magnitude of the probable error of a method. The estimate is derived from the fact that the average value obtained from replicate measurements of a single sample is a function of the errors of the individual determinations. That is, the average is defined according to

$$x_{av} = F(a_1, a_2, a_3, \ldots) = \frac{\Sigma_i a_i}{N} \qquad (6.1)$$

where the a_i are the results of the individual determinations.

But from Euler's theorem, which we state without proof, a small deviation in x_{av} may be written as

$$dx_{av} = \Sigma_i \frac{\delta F}{\delta a_i} da_i \qquad (6.2)$$

Equation 6.2 states that a small change (or error) in the average can be expressed as the sum of the individual errors in each of the determinations contributing to the average.

If we assume that the errors, $(\delta F/\delta a_i)da_i$, are all random, then, like the deviations about the mean of a normally distributed sample, we would expect them to sum to zero. So, in order to estimate the probable error in an average value, we also deal with the variance, i.e., the square of dx_{av}

$$dx_{av}{}^2 = \left(\Sigma_i \frac{\delta F}{\delta a_i} da_i\right)^2$$

Squaring dx_{av} yields

$$dx_{av}{}^2 = \Sigma_i \left(\frac{\delta F}{\delta a_i}\right)^2 da_i^2 + \text{(cross terms involving sums of products of } da_i)$$

But since we have assumed that the fluctuations are random, the cross terms should also be composed of about equally distributed negative and positive values and, thus, on average, sum to zero, leaving only the first term

$$dx_{av}^2 = \Sigma_i \left(\frac{\delta F}{\delta a_i}\right)^2 da_i^2 \qquad (6.3)$$

as the expression for the error in x_{av}. Equation 6.3 is the statement of the well known relationship describing the total error of any measurement. That is, it states that the error in an average value is equal to the square root of the sum of the squares of the errors of the individual determinations.

Returning to our definition in eq 6.1, and using the error function in eq. 6.3, we may now estimate the probable error in the average value. That is, taking partial derivatives of F in eq. 6.1, with respect to the a_i, we note that each is equal to 1/N. Furthermore, since we have assumed that the uncertainty in our replicate measurements should be normally distributed, the errors in each of the individual determinations should be about equal to the population standard deviation σ, i.e., $da_i \approx \sigma$. Substituting these values into eq. 6.3, we get the expected error in any set of replicates

$$dx_{av} = \left[\Sigma \left(\frac{1}{N} \right)^2 \sigma^2 \right]^{1/2} = \left(\frac{N\sigma^2}{N^2} \right)^{1/2} = \frac{\sigma}{N^{1/2}}$$

And, as with the population sample discussed before, the value of σ may be estimated by using the observed standard deviation of the set of measurements. Thus, the true value of x is given by

$$x = x_{av} \pm \frac{\sigma}{(N-1)^{1/2}}$$

where, again, we have substituted $(N - 1)$ for N in the denominator for the same reason as was mentioned before.

7

Physical Methods

SPECTROSCOPY

Introduction

Spectroscopy is a generic term that covers a diversity of analytical methods that are directed to quantitative chemical analysis as well as to investigations of molecular structure. The principle underlying all the methods, however, is the measurement of the intensity and/or the energy (wave length) of radiation that is either emitted, reflected, or absorbed by the substance being investigated, remembering that such absorptions and emissions are "quantized," i.e., they may be composed only of fixed increments of energy that are characteristic of allowable processes or "transitions" that atoms and molecules may undergo. That is, as has been discussed earlier, an electron may be promoted from a specific electron shell or energy level to another, but not in between. Stimulation of that promotion will always require the same quantum level of energy. Similarly, atomic vibrations in molecules are quantized, and only specific energies can be absorbed to excite a vibrating species to a higher energy level. Thus, the absorption "spectrum," i.e., the set of specific energies a particular analyte can absorb to cause one or more transitions, will always be composed of radiation of one or more wave lengths, each of which is associated with one of the transitions. A typical infrared absorption spectrum is shown in Figure 7.1.

One of the fortunate facts of nature is that the different kinds of transitions that atoms and molecules may undergo, e.g., electronic, vi-

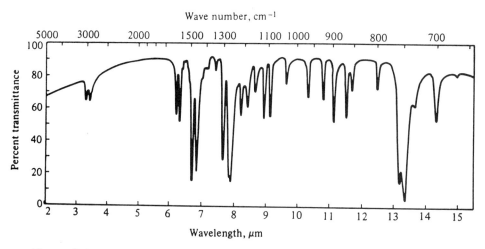

Figure 7.1   A typical infrared absorbtion spectrum, in this case for a compound of molecular formula $C_{13}H_{10}O$ encapsulated in a KBr pellet.

brational, rotational, etc., that lead to emission or absorption of energy, are associated with different regions of the spectrum. That is, molecular vibrations, ionization, fluorescence, etc. are, in general, stimulated by different spectral energies. Thus, the selection of the appropriate region of the spectrum allows for investigating the individual transitions or for quantitating specific analyte types. The several types of spectroscopy, therefore, may be distinguished by the region of the spectrum in which their measurements are made as well as by the kind of information being sought, as is illustrated in Figure 7.2.

As can be seen, spectroscopic methods range from highly energetic processes that stimulate changes in atomic nuclei to processes that only affect the magnetic orientation of electrons and protons. Neutron activation analysis, for example, makes use of high energy neutrons to transmute analyte atomic nuclei into excited radioactive states, and the energy and intensity of the decay products of the excited nuclei are used to quantitate the analyte.

Irradiation with rays of lower energy, like x-rays, knock electrons from inner electron shells out of the atom completely and, as the excited electrons return to their ground states, x-rays of somewhat lower energy are re-emitted. The intensity and frequency of such x-ray "fluorescence" may be related to the type and number of atoms involved.

At the still lower energies of flames or electric arcs, the outer electrons

Wave-length	Spectra Region	Transition Types	Types of Spectroscopy
25cm	Radio Waves	Proton & Electron Spin	Magnetic Resonance
0.04cm	Micro-waves		Microwave
25μm	Infrared Region	Molecular Rotations	Infrared
2.5μm		Molecular Vibrations	
8000Å			
4000Å	Visible Region	Valence	Visual and Ultraviolet
2000Å	Ultra-violet	Electron Transitions	Absorption & Emission
10Å	X-Ray Region	Inner Electron Transitions	X-ray Fluorescence
1Å	Gamma Rays	Nuclear Transitions	Neutron Activation Analysis

Figure 7.2   The spectral regions, the kinds of transitions excited by them, and the associated type of spectroscopy.

of the atoms may be excited and, upon return to ground state, will emit radiation in the ultraviolet and visible regions of the spectrum, the intensity and frequency of which can be used to quantitatively analyze a sample for a specific element.

In passing from very high to lower energies, the transitions that are affected change from those involving the atomic nucleus to those involving the inner electrons of the atom and, thence, to those involving the outer electrons. Continued reduction in radiative energy to the infrared and radio wave regions is then associated with molecular transitions and transitions in the magnetic orientations of electrons and protons.

Infrared radiation, for example, stimulates inter-atomic vibrations and rotations in molecules, and the still lower energies of the radio wave region of the spectrum stimulate electrons and certain protons to "flip" in their orientation in a magnetic field, giving rise to "magnetic resonance" analysis. Structural as well as quantitative information may be derived from infrared absorption and magnetic resonance spectra, incidentally, because of the influence molecular structure has on vibrational and rotational modes in molecules and on the energies required to effect electonic or protonic magnetic transitions.

Absorption Spectroscopy

Perhaps the most universally used of all analytical spectroscopic techniques, however, is absorption spectroscopy, the type of spectroscopy that deals with the absorption of radiation in the region of the spectrum extending from the far infrared to the ultraviolet. And it is this technique to which we now turn our attention because it is applicable to analysis of our subject elements. Indeed, it is the basis of a number of reference methods for carbon, nitrogen, and sulfur pollutants.

To carry out quantitative analysis based on absorption spectroscopy, one uses an absorption spectrophotometer, a simplified schematic of which is shown in Figure 7.3. Its essentials include a light source, a device that disperses the light into its colored* components, a mirror that may be used

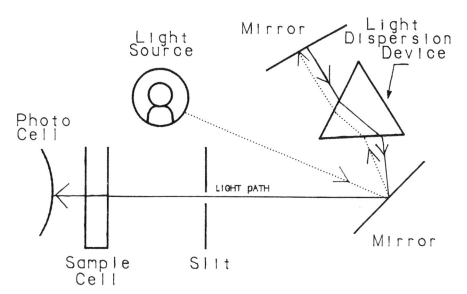

Figure 7.3   A simplified schematic of an absorption spectrophotometer.

---

*Colored components, in the context of this discussion, means the range of radiative energies contained in the output of the light source. For an incandescent light source, for example, the energy distribution could range from the near ultraviolet to the near infrared.

in concert with a "slit" to select a narrow range of colored components for transmission through the analyte sample, and a detector.

Light Sources

Measurements of absorption spectra generally require radiation sources that emit continuous bands of energy, i.e., emissions containing some fraction of all possible wave lengths within the working range selected. In that way, one is able to choose a specific wave length to effect a specific transition in the material being investigated. Such continuous emissions are generally characteristic of bodies that are heated to high temperatures.

   Thus, for operation in the visible part of the electomagnetic spectrum, incandescent bulbs are used as light sources; in the ultraviolet region, high temperature plasmas associated with gas discharges are used; and in the infrared region, heated ceramics are used.

   The spectral composition of several different light sources is illustrated in Figure 7.4. Note that a tungsten filament lamp, used in a visible

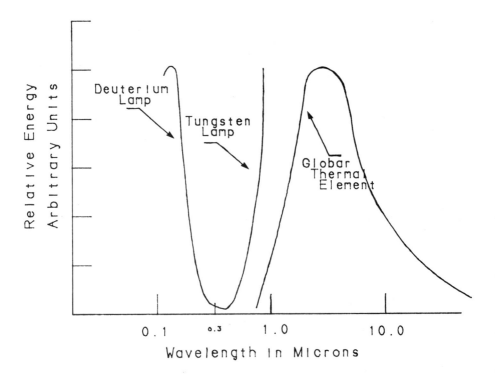

Figure 7.4   The emission spectra of several different kinds of light sources for use in absorption spectrometry.

spectrophotometer, emits a continuous band of increasingly intense wave lengths ranging from the near ultraviolet (≈ 400 Angstroms) to the near infrared (≈ 700 Angstroms). For the ultraviolet region, deuterium or hydrogen gas discharge lamps are generally used. The internal gases of these lamps are ionized by the energy absorbed as an electric current is passed through them. Upon return to ground state, the ionized gases emit radiation as a continuous band ranging in wave length from about 150 to 400 Angstoms. Finally, infrared light sources are, typically, heated bars of silicon carbide or coils of nichrome wire that radiate continuous energy bands with wave lengths that range from about 1 to 25 microns*.

Light Dispersion Devices

Prisms are frequently used as light dispersion devices in spectrophotometers. The action of a prism in dispersing light, as is well known, depends upon the velocity of light within the prism as compared to its velocity in a vacuum. The ratio of these two velocities is known as the index of refraction. Because of differences in indices of refraction between any two media, a wave front of monochromatic light from one, that is incident upon the other at some angle Φ ($\Phi < 90°$), will be bent as it enters the second medium. This is shown in Figure 7.5a.

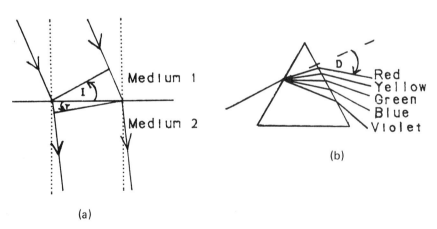

(a)

(b)

Figure 7.5   (a) Bending of a light wave as it is passed across the boundary between media of different refractive index; (b) the dispersion of white light into its colored components by a prism.

*A micron is 10^{-4} cm or 10000 Angstroms.

Now, in a vacuum, the velocity of light is independent of its wave length. But because of interaction with matter, the velocity of light in any other medium is wave length dependent. Thus, the index of refraction in any material is also wave length dependent. As a result, when white light strikes a prism surface, its components are bent at different angles that depend upon their wave length, and they are dispersed into the spectrum of which the white light is composed (Figure 7.5b).

Spectrophotometer prisms are made from materials with refractive indices and light absorption properties that will resolve broad band radiation in each of the regions of interest. Thus, for the visible region, ordinary glass may be used while in the ultraviolet, silica (SiO_2), in the form of fused quartz, is used. Infrared resolving prisms are made from ionic crystals like potassium or sodium chloride.

Diffraction gratings may also be used in spectrophotometers to resolve light into its components. Gratings are made of glass plates upon which a series of parallel lines have been scored in line densities of the order of several hundred to several thousand per cm. Light incident upon the surface of the grating passes between the lines. As it exits from the grating, it generates what are effectively independent light sources that emanate in phase from each of the line spacings. These independent light sources then interfere with one another as they spread out in circular wave fronts away from the grating, and the interference patterns produced resolve into the components of the source light.

Sample Cells

Sample cells are made from the same material classes as prisms in absorption spectrophotometers. Thus, for the visible region, ordinary glass cuvettes, usually with an optical path of one cm, are normally used; and fused quartz or silica cells are used for the ultraviolet regions. For the infrared, samples may be dispersed (or mulled) in an infrared transmitting fluid like mineral oil and then spread between to optically flat potassium chloride or sodium chloride windows. Alternatively, the sample may be intimately mixed with dry potassium chloride powder and pressed into the form of a pellet, which itself serves as the sample container. Gas samples for infrared analysis are analyzed in cells of 10 cm length, with infrared transmitting windows of KCl or NaCl sealed to their ends, and gas entry and exit ports fixed to their walls. For very dilute gas analysis, long path cells are available in which internal mirrors are used to reflect the IR beam back and forth and increase the absorption path length.

Detectors

Practically all modern visible and ultraviolet spectrophotometers make use of photomultiplier tubes as detectors. A photomultiplier is essentially a vacuum tube containing a battery of electrodes, called dynodes, electrically arranged in series with a high potential between each pair (Figure 7.6). The primary electrode, the photo-cathode, is coated with a light

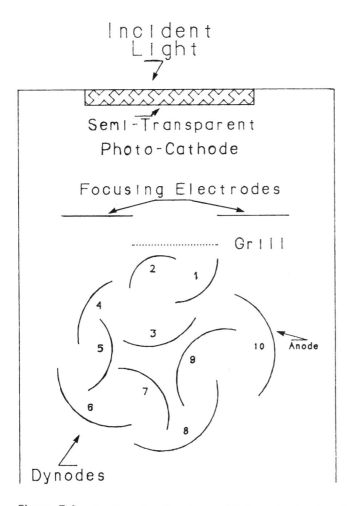

Figure 7.6 A schematic of a photomultiplier tube showing the arrangement of the dynodes.

sensitive material like cesium, which will emit photoelectrons if it is irradiated with light ranging in wave length from about 2000 to about 10,000 Angstroms. The emitted photoelectrons are then accelerated to the first dynode by the high potential field, where each one stimulates the emission of several other electrons as it strikes. The resulting cascade, as electrons from each dynode are accelerated toward the next, causing still more emissions, provides an amplified output related to the intensity of the original light signal. The amplification factor is determined by the number of dynodes in the device. Thus, for example, if each of the photoelectrons produced by the primary electrode produced four secondary electrons at the first dynode, and each secondary electron produced four others in the cascade, a tube with ten dynodes would produce an amplification factor of 4^{10} or about one million. As a result, photomultipliers, are extremely sensitive and efficient devices for measuring light.

There are a number of different types of detectors that may be employed in infrared spectrophotometers. Included are photoconductive devices, thermocouples, and bolometers. Photoconductors are semi-conductor materials whose valence electrons are easily promoted by absorption of light to the higher energy levels associated with electrical conductivity. Most photoconductive semi-conductors are thus stimulated by energies starting in the near infrared and extending out to several microns. For infrared spectrometers operating in regions beyond a few microns, however, thermal detectors like themocouples and bolometers must be used. Thermocouple devices are made of junctions of dissimilar metals or alloys which, when heated by infrared radiation, generate a voltage that may be amplified and recorded. A bolometer is basically a resistor with a very high temperature coefficient of resistivity that is incorporated into a bridge circuit. When it is heated, its change in resistance upsets the balance of the bridge circuit and allows current to flow through a detector.

Quantitation

Quantitation in absorption spectroscopy is based on the Beer-Lambert law. The law was originally formulated by Lambert in the middle 18th century when he observed that the reduction in the intensity of light as it passed through a thin film of material was dependent upon the wave length of the light and the nature of the film material, and was proportional to the thickness of the film. In later years, in the middle of the 19th century, Beer modified Lambert's formulation by noting that one could regard the thin film as a container containing a fixed concentration of light absorbing nuclei. Thus, the decrease in intensity of light passing through a film of thickness dx could be written as

$$\frac{dI}{dx} = -\epsilon cI \qquad (7.1)$$

with I being the intensity, c the concentration of light aborbing nuclei, and ϵ a constant of proportionality. Integration of eq. 7.1 then yields the expression of the Beer-Lambert law

$$\log \frac{I}{I_0} = -(\epsilon)(c)(l)$$

where l is the thickness of the absorption element being investigated. The ratio I/I_0 is known as the sample *transmittance* while the *absorbance*, A, is defined as equal to the negative of the logarithm of the transmittance or

$$A = -\log\left(\frac{I}{I_0}\right) = (c)(l)(\epsilon)$$

and the proportionality constant ϵ is called the molar absorptivity when the analyte concentration is expressed as solution molarity.*

Use of a Spectrophotometer

A spectrophotometer may be used to serve two purposes. In one, wave lengths of maximum sample absorbance. i.e., the absorption spectrum of the unknown in the sample cell, may be determined by rotating the spectrophotometer prism (Figure 7.3), irradiating the sample with the selected spectral components of the source light, and measuring the trans-missions of each. The slit is used to limit the band width, i.e., the wave length range of the source light's components that may be transmitted through the sample.

In quantitative analysis, however, one usually already knows the ab-sorption spectrum of the unknown, which may be the analyte itself or some chemical derivative of it with a detectable spectrum. One of the more intense absorption peaks of the analyte is pre-selected, and a blank, i.e., a sample expected to have zero concentration of analyte or 100% transmittance is used to "zero" the instrument. Then one measures the

---

*Molarity is defined as the number of formula weights of analyte per liter of water.

transmittance of a number of known levels of analyte to generate a calibration curve. The transmittance of the unknown is finally measured and interpolated onto the calibration curve for quantitation.

Specificity and Sensitivity of Spectrophotometric Measurements

The specificity of spectrophotometric measurements is limited by the extent to which other substances in the sample absorb energy in the same region as the analyte. To minimize the problem, it is frequently necessary to chemically modify the analyte so as to form a new material that exhibits more unique absorption bands or to chemically modify the interfering material to make it absorb in another region of the spectrum.

Sensitivity is determined, in large measure, by the intensity of the absorption band, so in developing new methods, preliminary activity is directed to identifying the most intense bands or to chemical modifications that will yield intense absorption bands. Smythe [1977, p. 682] has suggested that visible and ultraviolet spectrophotometric determinations of analyte levels in the 10-100 ppm range may be made with a precision of 1-5% while samples with concentrations of analytes in the 0.005-0.1 ppm range can be measured with a precision of 5-10%.

Prism and grating infrared spectrophotometers are somewhat less sensitive and yield precision of the order of 10% for sample concentrations of 10-100 ppm. It is perhaps because of its limited sensitivity that infrared spectroscopy is not in common use for routine environmental analysis, nor is it applied in any reference methods. It is useful, however, for identifying unknown contaminants in air and water.

Error in Spectrophotometric Measurements

To minimize error and enhance precision in spectrophotometric measurements, it is always recommended that the analyzed sample be diluted (or concentrated) so as to yield transmittance values in the mid-range, i.e., between about 30 and 60%. The reason for this stems from the logarithmic nature of the Beer-Lambert law. This is illustrated in Figure 7.7 in which transmittance is plotted as a function of concentration. As can be seen, small errors in reading transmittances lead to larger errors at the extremes of transmittance values than they do in the midrange. That is, at low transmittance values, small reading errors lead to large absolute concentration errors. At the other end of the scale, small reading errors lead to concentration errors that are large relative to the absolute concentration levels being determined.

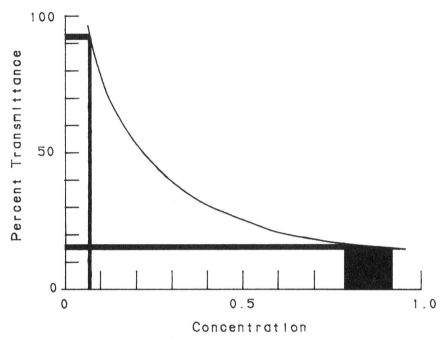

Figure 7.7   An illustration of the error possibilities resulting from making spectral absorption measurements at the extremes of transmittance. Note that at high
transmittance levels, the relative error is large. At low transmittance values, the
absolute error is large. (see text).

Colorimeters

As a last point in our discussion of absorption spectroscopy, it is necessary
to mention a class of instruments called filter colorimeters. These instruments, which are generally inexpensive, may also be used effectively for
quantitative analysis in the visible region of the spectrum. In colorimeters,
the light dispersion components of the spectrophotometer are replaced
with colored filters. Thus, if one knows the absorption spectrum of the
analyte, one can select a colored filter that transmits only one of the
absorption peaks and measure, as in a spectrophotometer, the ratio of
unknown to known absorbance to conduct quantitative analysis.

Spectrophotometric Analysis of Carbon, Nitrogen, and Sulfur

As was mentioned earlier, spectrophotometric measurements serve as
the basis of several reference methods for carbon, nitrogen, and sulfur
pollutants in air [40 CFR, Part 50] and water [40 CFR, Part 136]. We
present here some of the highlights of those procedures.

Carbon   With respect to carbon, analysis of specific compounds is specified by methods to be discussed later. Oxidizables in water, however, which are primarily organic carbon compounds, are estimated by measuring COD (Chemical Oxygen Demand—see Chapter 4). COD is determined by boiling a water sample together with potassium dichromate and sulfuric acid. Organics in the water are oxidized to carbon dioxide, and the dichromate ion ($Cr_2O_7^{2-}$) is reduced to Cr^{3+}, an ion with a strong absorption band at 585 nanometers (5850 Angstroms). Calibration of the spectrophotometer or colorimeter used to measure the absorbance of the Cr^{3+} ion in the digested sample is carried out by preparing standards containing known concentrations of oxalic acid. The standard solutions are digested similarly to unknowns, and their absorbances at 585 nm are measured to develop the calibration curves. The method is capable of measuring COD levels between about 10 and 100 ppm. Samples with anticipated higher values are diluted previous to analysis.

Nitrogen   There are a number of spectrophotometric reference and equivalent methods for nitrogen pollutants in air and water, the details of which are available from the EPA. By way of example, however, we present the essentials of several of them here.

Ammonia in water is determined by reacting the sample with a solution of potassium mercury iodide (Nessler's reagent). The resulting brown solution is quantitated with a filter colorimeter using filters transmitting in the neighborhood of 425 nm. Ammonia may also be determined in water with a solution of sodium phenylate and sodium hypochlorite.* The reaction results in the formation of a colored solution of indophenol blue, the concentration of which is proportional to the ammonia. Quantitation may be accomplished by measuring the absorption of the indophenol blue at 630-660 nm. Full descriptions of these methods, incidentally, may also be found in ASTM Standard D 1426.

Nitrate† in water is determined by first reducing it to nitrite by passing the sample through a column containing copper coated cadmium granules (ASTM Standard D 3867). The nitrite may then be quantitated by reacting it with an aromatic amine, in particular, sulfanilamide, in a so-called diazotization reaction, and thence with N-(1-Naphthyl)-Ethylenediamine Dihydrochloride to form an intensely colored azo dye. The absorbance of the dye product is then measured spectrophotometrically at 543 nm. Nitrites, of course, may be determined similarly but with the elimination

*One of EPA's recommended methods for determination of the constituents of acid rains [Operations and Maintenance Manual 1986, Sect. 4, p. 9].
†One of EPA's recommended methods for differentiation of the constituents of acid rains [Operations & Maintenance Manual 1986, Sect. 4, p. 5].

of the cadmium reduction step. The applicable range of this determination is from about 0.05 to 1 ppm as nitrogen.

Nitrogen dioxide in air may be determined similarly and is the basis of both an EPA equivalent method[‡] and ASTM Standard D 1607. In the latter, a known volume of air is bubbled through an "absorbing" solution containing sulfanilic acid and N-(1-naphthyl)ethylenediamine dihydrochloride, forming an azo dye that is measured at 550 nm. Standards for the method are based on the use of sodium nitrite solutions, 0.72 moles of which having been found to produce color corresponding to one mole of nitrogen dioxide. The recommended range of this determination is 10 to 400 μg NO_2/m^3. For the EPA reference method, the absorbing solution is a solution of sodium hydroxide and sodium arsenite, and color is developed with sulfanilamide and N-(1-napthyl)-ethylenediamine dihydrochloride to form the azo dye as described above.

Sulfur The primary reference method for sulfur dioxide in air is a spectrophotometric procedure based on the the formation of the intensely bright red-violet compound, p-rosaniline methyl sulfonic acid, which is formed when bisulfite ion is reacted with p-rosaniline[§] dye and formaldehyde [40 CFR Part 50.12, Appendix A]. The procedure is presented in detail in ASTM Standard D 2914. In brief, however, air samples are aspirated through an "absorbing" reagent composed of a solution of mercuric chloride and potassium chloride adjusted to a pH of about 4 with sodium hydroxide. This absorbing reagent reacts with SO_2 to form the disulfitomercurate ion $\{Hg(SO_3)_2^{2-}\}$, which then reacts with the p-rosaniline and formaldehyde. The product solutions are measured at 548 nm, and calibration standards are prepared from solutions of sodium sulfite. The range of the method is from approximately 25 μg SO_2/m_3 to 1000 μg SO_2/m_3. The detection limit is of the order of 4-10 μg SO_2/m_3, depending upon how long one aspirates the air sample through the absorbing solution.

The spectrophotometric reference method for sulfide in water is based on the formation of methylene blue dye from the reactions of sulfide ion, ferric chloride, and p-aminodimethylaniline [Standard Methods 1981, Sect. 427C and 40 CFR Part 136.3, 249]. Standards are prepared from solutions of known concentrations of sodium sulfide trihydrate, and absorbance is measured at 600 nm. The minimum detectable concentration of sulfide, using the methylene blue method is 50 μg/l as sulfide.

[‡]EPA Designated Equivalent Method No. EQN-1277-026.
[§](4-amino-3-methylphenyl)bis(4-aminophenyl)methanol.

Non-Dispersive Infrared Absorption Spectroscopy

Principles

Non-dispersive infrared absorption (NDIR) is a special application of infrared spectroscopy. The method is best explained by reference to Figure 7.8, which is a schematic of a Luft type NDIR cell. In the Luft cell, light from a broad band infrared source is directed by means of a pair of mirrors through the ends of twin gas chambers, one serving as a reference chamber and the second a sample chamber. The ends of the gas chambers are sealed with infrared transmitting windows. Between the mirrors and the gas chamber entry windows is a rotating mechanical shutter that alternately blocks and transmits the radiation that may enter the chambers. At the opposite ends of the chambers is a two compartment detector cell fitted with windows matching those of the gas chambers. The compartments of the detector cell are separated by a thin metal membrane which serves as one plate of a differential capacitor, and the compartments are filled with the analyte gas. A suitable analyte-free carrier gas is circulated from the reference chamber to the sample chamber.

In using the Luft cell, advantage is taken of the increase in pressure the gas in the compartments of the detector section experiences as it absorbs radiation and heats. If one of the compartments should heat more than the other, the thin metal membrane is displaced because of the pressure difference. But as long as no analyte is in the sample chamber,

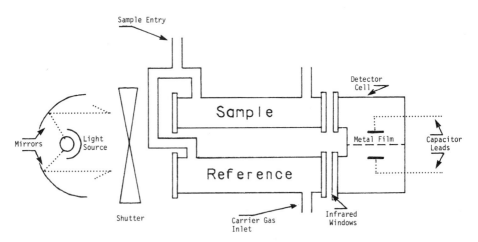

Figure 7.8   A schematic of a Luft type nondispersive infrared cell.

both compartments of the detector cell experience the same amount of radiation and there is no pressure differential. However, if a sample of analyte gas is injected into the sample cell, it will absorb some of the radiation passing through the chamber, and less will be transmitted into the sample side of the detector section. Thus, the gas in that section will cool relative to the reference side and the metal film will be displaced. Because the mechanical shutter blocks the radiation periodically, the pressure differential will also alternately rise and fall yielding a periodicity to the metal film displacement and a periodic change in the differential capacitance. The last can be detected and amplified and is directly proportional to the amount of analyte gas in the sample cell.

Luft cells are efficient because all the effective absorption bands of the analyte contribute to the signal. This is in contrast to dispersive infrared instruments in which only one specific line or band is selected for analysis. Luft cells also exhibit selectivity because interfering gases will, more than likely, have spectra that are quite different than the analyte gas (and the gas in the detector). That is, the spectral peaks of an interfering gas will not match those of the detector gas. Thus, the energy the interfering gas absorbs will have no effect on the temperature of the detector gas.

Applications

NDIR may be used for analysis of a number of infrared absorbing gases like carbon dioxide, carbon monoxide, nitric oxide, and sulfur dioxide. In fact, NDIR detection is used in EPA's designated reference method for carbon monoxide in air [40 CFR, Part 50, Appendix C]. It is also one of the prescribed detection methods for measurement of organics in water [40 CFR, Part 136.3] and is the basis of ASTM Standard D 2579 A for determination of the total organic content of water (see Chapter 10).

Fourier Transform Infrared Spectrosocopy

Principles

Fourier transform spectroscopic methods are relatively new to analytical chemistry, although they have been used by astronomers to study radiation emitted from the stars for a long time. In contrast to dispersive spectroscopy, in which emitted and absorbed radiation is analyzed by means of prisms or diffraction gratings, Fourier transform instruments make use of "interferometers." An interferometer is a device that splits a beam of radiation into two and then recombines them in such a way as to develop "interference" patterns which contain, as we shall show, accessible information about the properties of the source radiation.

A schematic of an interferometer is shown in Figure 7.9b. It is composed of two mirrors oriented perpendicular to one another, one fixed in position, and the other movable. Between the mirrors is a beam splitter, which might be a half silvered mirror or mylar film, oriented in such a way that half the light from a radiation source is transmitted to the movable mirror, and the other half is reflected to the fixed mirror.

Now, if the radiation source is monochromatic, and if the distances between the beam splitter and the movable and fixed mirrors are exactly equal, then the radiation reflected from the mirrors back to the beam splitter will be in phase, i.e., the crests and troughs of the reflected light waves will match, and the two beams will sum to the intensity of the source radiation (Figure 7.10a); half will then return to the source and half will be reflected through the sample to the detector. However, if the movable mirror is displaced, say a distance equal to half a wave length of the source radiation, the beams returning from the two mirrors to the splitter will be 180° out of phase and will destructively interfere with one

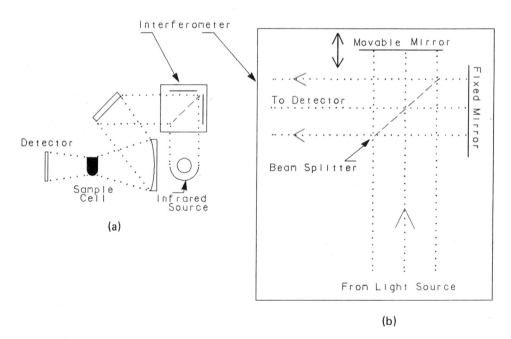

(b)

Figure 7.9   (a) A simplified schematic of a Fourier Transform Infrared Absorption Spectrometer; (b) a schematic of an interferometer used in an FTIR.

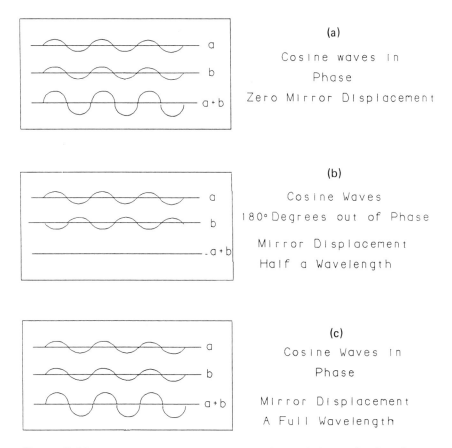

Figure 7.10   An illustration of the constructive and destructive interferences that appear at the beam splitter in an interferometer, which is caused by displacement of the mirror (see text).

another (Figure 7.10b), and no light will return to the source or be reflected to the detector. If the mirror is displaced another half wave length, the two reflected beams will again constructively interfere (Figure 7.10c), and a signal will again be received at the detector. It can be seen, therefore, that as the movable mirror is displaced at constant velocity, the intensity of the light striking the detector will vary sinusoidally as the displacement distance extends over multiple integrals of the wave length. Some reflection will reveal that the signal at the detector will be given, for monochromatic source radiation, by

$$I(x) = 0.5(I_l)\left(1 + \frac{\cos 2\pi x}{l}\right)$$

or

$$I(x) = 0.5(I_l) + 0.5(I_l) \cos \frac{2\pi x}{l}$$

where $I(x)$ is the signal intensity as a function of mirror displacement, x is the displacement, and I_l is proportional to the intensity of a radiation source of wave length l.

In infrared Fourier transform instruments (FTIR), the mirror is moved at constant velocity. Thus, we may substitute $v_0 t$ for x, where v_0 is the mirror's velocity and t is time

$$I(x) = 0.5(I_l) + 0.5(I_l) \cos \frac{2\pi v_0 t}{l} \qquad (7.2)$$

That is, in an FTIR spectrometer, the intensity of the signal at the detector is a periodic function of time and also depends upon the *wave length* of the source radiation.

As can be seen, however, only the second term in eq. 7.2 contains information about both the intensity and the wave length of the source radiation. So, the expression

$$I(t) = 0.5(I_l) \cos \frac{2\pi v_0 t}{l} \qquad (7.3)$$

which contains only the second term of eq. 7.2, is the one that is used to analyze the interference data appearing at the detector. The expression for $I(t)$ is called an *interferogram*.

Equation 7.3, of course, represents only a monochromatic radiation source. For more complex sources, say a beam of two different wave lengths, the detector would see a signal composed of the sum of the two cosine waves generated by the "bi-chromatic" source. That is

$$I(t) = 0.5\left[(I_1) \cos \frac{2\pi v_0 t}{l_1} + (I_2) \cos \frac{2\pi v_0 t}{l_2}\right]$$

And extending to polychromatic radiation, the signal arriving at the detector would be the sum of all the cosine waves generated by the individual components of the source. Thus, the interferogram generated by poly-chromatic radiation would be given by an equation of the form

$$I(t) = K\Sigma_i(I_i) \cos \frac{2\pi v_0 t}{l_i} \tag{7.4}$$

Having this complex interferogram, which we see contains information about the intensity and wave length of all the components of the source radiation, we are left with the question of how to extract the information from it. The approach used to do so is also the source of the method's name. That is, one takes the Fourier transform of eq. 7.4 to develop the spectrum that is contained within the interferogram.

To explain the meaning of a Fourier transform, one must recognize that any curve, even, for example, a square or triangular pulse, may be represented by the sum of appropriate combinations of cosine waves of different frequencies and amplitudes. Thus, one can synthesize any curve shape by choosing the proper combination of cosine waves and algebraic-ally summing their amplitudes as they sweep out along the abscissa. The Fourier transform of the curve is the inverse of its synthesis; it is the mathematical manipulation required to determine the composition of the curve's component cosine waves, i.e., its spectrum.

The formal mathematics of Fourier transforms is beyond the scope of this discussion. However, we can outline the ideas and manipulations involved in extracting the spectral information from complex curves like interferograms.

Basic to their analysis is the well known trigonometric identity for the product of two cosines

$$\cos(a) \cos(b) = \frac{1}{2}[\cos(a + b) + \cos(a - b)]$$

The analysis proceeds by multiplying each term on the right hand side of eq. 7.4 by a cosine term containing a single wave length, as for example, $\cos(2\pi v_0 t/l_2)$, and invoking the identity. That is,

$$\Sigma_i(I_i) \cos \frac{2\pi v_0 t}{l_i} \cos \frac{2\pi v_0 t}{l_2}$$

$$= \Sigma_i(I_i) \left[\cos 2\pi v_0 t \left(\frac{1}{l_i} + \frac{1}{l_2} \right) + \cos 2\pi v_0 t \left(\frac{1}{l_i} - \frac{1}{l_2} \right) \right] \tag{7.5}$$

Then eq. 7.5 is integrated over the time period of the measurement giving

$$\int \Sigma_i(I_i) \cos \frac{2\pi v_0 t}{l_i} \cos \frac{2\pi v_0 t}{l_2} dt =$$

$$\int \Sigma_i(I_i) \left[\cos 2\pi v_0 t \left(\frac{1}{l_i} + \frac{1}{l_2} \right) + \cos 2\pi v_0 t \left(\frac{1}{l_i} - \frac{1}{l_2} \right) \right] dt \qquad (7.6)$$

But the integration of the cosine terms on the right hand side of eq. 7.6 over time is, in fact, a calculation of the area they sweep out in time. Since cosine functions oscillate symetrically about the abscissa, they sweep positive and negative areas equally. Therefore, on average, each of the terms of eq. 7.6 will yield a net area of zero, except for one in which $l_i = l_2$. That is, the only non-zero result of the integration will be

$$\int I_2 \cos 2\pi v_0 t \left(\frac{1}{l_2} - \frac{1}{l_2} \right) dt = \int I_2 \cos(0) \, dt = \int I_2 \, dt$$

This property of the integrals of cosine products thus provides a way of interrogating the interferogram to determine whether or not it contains a particular wave length. One samples the interferogram within the time interval of the measurement and then multiplies each sampling point by the cosine of a selected wave length. Then, computation of the area swept out by the resulting cosine product function will yield a zero value if the selected wave length is not a component of the interferogram and a non-zero value if it is. One checks sequentially for all possible values of the wave length to isolate the spectrum contained in the interferogram.

Carrying out these kinds of manipulations on interferograms, i.e., taking their Fourier transforms, yields the kinds of results illustrated in Figure 7.11. In Curve "a," the amplitude of a cosine function is plotted versus time. The Fourier transform of "a" is shown to the right and is seen to be simply a vertical line on the abscissa at the value of the wave length of "a," with an amplitude proportional to that of the original function. The transform is the spectrum of Curve "a."

For a function that is a sum of two cosine waves, the Fourier transform would appear as two vertical lines, each with amplitude proportional to its original cosine wave. Thus, an interferogram resulting from two monochromatic light sources might appear as in Curve "a + b," and its Fourier transform would be as on its right. The bottom curves are the interferogram representing the sum of three cosine waves and its associated Fourier transform.

In general then, it can be seen that the Fourier transform of an interferogram generated by a polychromatic radiation source is, indeed,

Cosine Waves   Fourier Transform

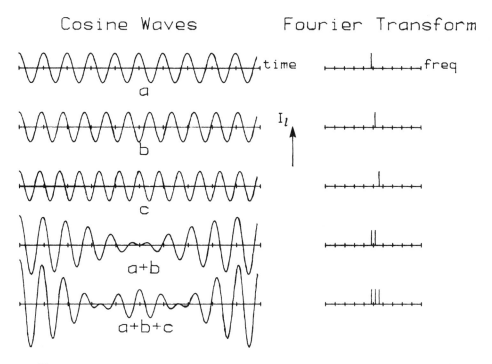

Figure 7.11 Monochromatic and polychromatic interferograms and their associated Fourier transforms. Wave intensity is given by I_l. In the cosine waves, the abscissa is time while in the transforms, the abscissa is wave length.

the spectrum of the source. In carrying out an FTIR analysis, the sample is placed in the light path between the interferometer and the detector (Figure 7.9a). The light exiting the sample, then, is an interferogram that contains intensity and wave length information about the source radiation minus any components absorbed by the sample. Its Fourier transform, therefore, is the transmission (or absorption) spectrum of the sample.

FTIR Advantages

As was mentioned above, interferometry has long been used by astronomers to analyze radiation from heavenly bodies, especially when the emissions were of low intensity. This is because interferometric measurements provide improvements in signal to noise ratio and, thus, in sensitivity and resolution compared to dispersive techniques. The nature of the optical systems in interferometers allows them to transmit light with efficiencies as high as 200 times that of dispersion spectrometers [Green and Reedy 1978, 21] and, like NDIR, all of the source light is used for analysis, in contrast to the one small spectral region defined by the slit and prism (or grating) of a dispersion spectrometer.

The delay in applying Fourier transform methods to analytical chemistry was primarily due to the obviously complicated and time consuming numerical manipulations required to carry out the transformations. With the advent of inexpensive digital computers, however, along with the invention of an algorithm especially suited to computing Fourier transforms, called the Fast Fourier Transform, the method has seen increasing and even revolutionary use in analytical chemistry.

Applications

Though it has not, as of the date of this writing, been incorporated into any of the U.S. regulatory agencies' reference or equivalent methods, FT-IR's sensitivity and spectral resolution capabilities suggest that its application to environmental analysis is inevitable. Because of its enhanced resolution, it is particularly useful in the far infrared region; because it is in that region of the spectrum that very fine spectral details occur that "fingerprint" specific compounds. As a result, and as will be seen in our later discussions on chromatography, FTIR is gaining increasing use for detection and analysis in the so-called "hyphenated" analytical method called Gas Chromatography-FTIR.

Reflectance Spectrometry

Principles

Polychromatic radiation incident on a surface will be partly absorbed and partly reflected. In reflectance spectrometry, the reflected light is analyzed. The chemical nature of the surface, of course, determines which of the wave lengths of the source radiation are reflected or absorbed, and the concentration of the absorbing constituent, as in absorption spectrometry, determines the absorption intensity.

In considering analysis of reflected light, a distinction must be made between two types, specular and diffuse. Pure specular reflection is the kind one gets from a mirror. The angle of incidence is equal to the angle of reflection as is shown in Figure 7.12a. On the other hand, a diffuse reflector as, for example, a piece of matte white paper, with incident radiation normal to its surface, will reflect light more or less equally at all angles within the solid angle represented by the hemisphere shown in

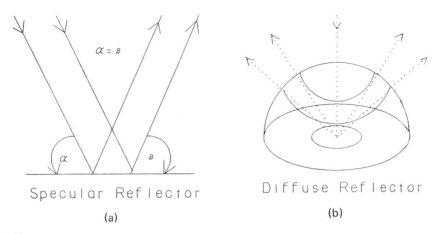

Specular Reflector
(a)

Diffuse Reflector
(b)

Figure 7.12   (a) The angular dependence of specular reflectance; b) the geometrically symetric distribution of reflected light rays from a diffuse reflector.

Figure 7.12b; a perfectly diffuse reflector will appear uniformly bright, regardless of the angle of observation.

If one were to layer a solution of known thickness on the surface of a pure specular reflector, one could analzye the layer by application of standard absorption spectrometric methods, treating the layer of material as the sample cell. Analysis of diffuse reflectance, however, requires a somewhat different approach than straightforward application of the Beer-Lambert law to a thin sample. It is based on a reflectance model developed in the early thirties by Kubelka and Munk [Kortum 1969, 106], which is shown in Figure 7.13. The illustration is that of a reflecting medium with a backing material of reflectance R_g. When light of intensity I_+ passes downward through the medium, some of it is absorbed and some of it is scattered (see Turbidimetry and Nephelometry section below). Light that is not absorbed ultimately reaches the backing layer where it is reflected back up through the medium with intensity I_-. As it moves up, more of it is absorbed and more is scattered, with light scattered from the downward directed radiation adding to the intensity of that directed up and vice-versa.

The equations describing the changes in intensity in the up and down directions are

$$dI_+ = -(S + K)I_+ \, dx + SI_- \, dx$$

and   (7.7)

$$dI_- = -(S + K)I_- \, dx + SI_+ \, dx$$

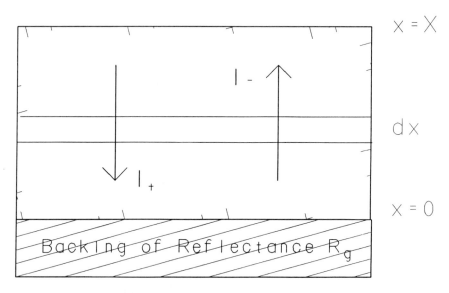

Figure 7.13   The Kubelka-Monk model for an infinitely thick diffuse reflector.

where S is a scattering coefficient that depends upon the texture of the reflecting medium, and K is an absorption coefficient that is proportional to the concentration of light absorbing species in the medium. Thus,

$$K = \alpha C$$

with α being an absorption coefficient analogous to the molar absorbance of the Beer-Lambert law.

Equations 7.7 may be integrated to yield

$$R = \frac{\left(\dfrac{1}{R_\infty}\right)(R_g - R_\infty) - \left(R_g - \dfrac{1}{R_\infty}\right) R_\infty \exp\left[Sx \left(\dfrac{1}{R_\infty} - R_\infty\right)\right]}{(R_g - R_\infty) - \left(R_g - \dfrac{1}{R_\infty}\right)\exp\left[Sx \left(\dfrac{1}{R_\infty} - R_\infty\right)\right]} \tag{7.8}$$

where R is the reflectance from the surface of a finitely thick reflecting medium and R_∞ is the reflectance of the infinitely thick medium. But it can also be shown [Kortum 1969, 110] that for a diffuse reflector of infinite thickness, where R_∞ is effectively zero and $R = R_\infty$, that R_∞ will be a function of (K/S) only, expressed in the form

$$\frac{(1 - R_\infty)^2}{2R_\infty} = \frac{\alpha C}{S} \tag{7.9}$$

It is again one of those fortunes of nature, however, that an effectively infinitely thick diffuse reflector is not so thick at all. In fact, a paper reflector of thickness of the order of 1-3mm (about the thickness of a typical test strip used for reflectance analysis) behaves as if it were infinitely thick. Thus, eq. 7.9 represents a working analogy to the Beer-Lambert law that may be applied to reflectance spectrometric analysis.

Reflectance Measurement

Two approaches to measurement of diffuse reflection are illustrated in Figure 7.14. The more efficient system is illustrated in Figure 7.14a. It makes use of an integrating sphere, a closed sphere with a highly reflective diffuse interior surface into which one places the sample to be analyzed. The sample is irradiated with light of appropriate wave length for the analysis. Then, any light that is not immediately reflected to the detector, is multiply reflected about the sphere until it does reach the detector or is absorbed by the sphere wall. In this way, nearly all of the light diffusely refected from the analyzed surface may be collected for measurement.

   Figure 7.14b is the simpler but less efficient method. One simply

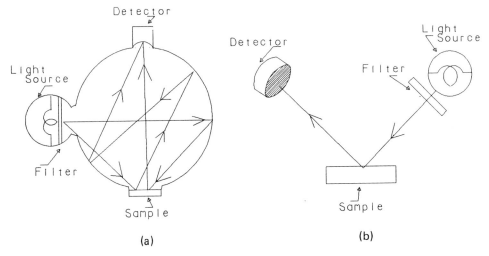

Figure 7.14   (a) A schematic of an integrating sphere used for measuring the intensity of diffusely reflected light; b) an alternative, but less efficient, method for measuring the intensity of diffusely reflected light.

lights the sample surface from some arbitrary incident angle and measures the reflectance with a detector oriented at some other angle. This method takes advantage of the fact that diffuse reflectors exhibit equal brightness regardless of viewing angle. But, as can be seen, the light collection efficiency is limited, because only the light that is reflected from the surface in the solid angle that intersects the detector is measured. All the rest is lost.

Error in Reflectance Measurement

Like standard absorption spectroscopy, measurements in the midrange of reflectance intensities yield the highest precision. This can be shown by differentiating eq. 7.9 to obtain the relative error in concentration. That is

$$\frac{dC}{C} \cdot (100) = \text{Relative Error} = \frac{R_x + 1}{(R_x - 1)R_x} dR_x$$

If one assumes a reading error of 0.3% in reflectance ($dR_x = 0.3$), which is a reasonable measurement expectation, and then plots relative error

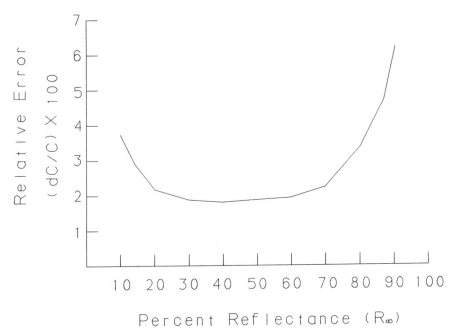

Figure 7.15   Relative error in diffuse reflectance measurement.

versus percent reflectance, as in Figure 7.15, one can see that the relative error rises exponentially at the extremes of the reflectance values.

Applications

Reflectance spectrometry serves as the basis of a large number of medical diagnostic tests. Perhaps more familiar, however, is its implicit use for testing the acidity of materials with graded pH papers, where the eye serves as a color analyzer analogous to a spectrometer. Its pertinence to our discussion is its application to the determination of hydrogen sulfide in the atmosphere (ASTM D 4323-84). The gas is measured by means of a paper strip which is impregnated with lead acetate and exposed to ambient air. Hydrogen sulfide reacts with the lead acetate to form black lead sulfide. In the ASTM method, measurement of the rate of formation of the black color is made with the use of the simpler form of reflectance spectrometer discussed above. The rate is then related to the concentration of the gas in the atmosphere. The method is sufficiently sensitive to detect volumes of H_2S in air as low as one part per billion.

TURBIDIMETRY AND NEPHELOMETRY

Principles

Turbidimetry and nephelometry are analytical techniques that are based on the light scattering properties of suspended solids. The two techniques differ only in the manner in which the scattered light is observed. In turbidimetry, measurement is made of light that is scattered in a forward direction by the suspended matter. In nephelometry, light that is scattered at a 90° angle from the direction of the source radiation is measured. In both cases, the intensity of the signal reaching the detector may be related to the concentration of the suspension.

Light scattering phenomena result from the interactions of electomagnetic radiation with the electric charges of matter. That is, all photons incident upon matter can excite oscillations between electrons and atomic nuclei and between atoms in molecules. Quantum mechanical restrictions, of course, permit only specific transitions to occur. So, in general, the incident energy that gives rise to the oscillations is immediately re-radiated as the excited entities relax to their original energy levels. The re-radiated energy, however, occurs in all directions, giving rise to what is called elastic scattering.*

---

*It is possible for some of the incident radiation to be absorbed. This can occur when a molecular vibration is excited by the incident photon and relaxes to an allowable state higher

There are two kinds of elastic scattering, Raleigh scattering from suspended particles that are much smaller ($<10\%$) than the wave length of the incident radiation, and large particle scattering where the particle sizes are about 10% to 150% of the wave length of the incident radiation.† Raleigh scattering produces a relatively symetrical pattern of radiation. That is, an observer viewing the scattered radiation, would see little difference as a function of viewing angle. On the other hand, in the case of large particle scattering, there is a greater concentration of scattered radiation in the forward or incident direction.

The differences in scattering properties between large and small particles are a consequence of light interference phenomena. Particles much smaller than the wave length of the incident radiation behave like point sources of scattered light. Larger particles, however, may have multiple points from which radiation is re-emitted. The increased concentration in the forward direction can be shown to be a consequence of interference among the light waves radiating from those multiple points [Olsen 1975, 26].

Thus, particle size plays an important role in turbidimetric and nephelometric analysis. For nephelometric measurements, it is preferable to have suspended particles small so as to avoid concentration of forward scattered light and, thereby, maximize the intensity of the signal seen at 90°. In turbidimetry, small particles are also preferred because measurement is based on the total amount of incident light that is excluded by the sample, regardless of the exclusion mechanism. Too much forward scattering, as well as multiple reflections from large particles, adds nonlinear factors to the relationship between light attenuation and concentration.

Instrumentation and Measurement

For turbidimetric measurements, one may use any spectrophotometer or colorimeter. Although the net transmittance of light through the suspension can be shown to be mathematically identical to the Beer-Lambert law, i.e., $\log(I_0/I) = kC$ [Olsen 1975, 455], measurements are usually made by interprolating the transmittance of an unknown within a calibration

---

in energy than the one it was in before interaction with the photon. The scattered radiation then will be lowered in energy by the difference between the two vibrational states of the molecule. This phenomenon is called inelastic or Raman scattering.

†Particles that are much larger than the wave length of the incident radiation reflect rather than scatter the light.

curve constructed from known suspensions prepared identically. This is because the proportionality constant for turbidimetry is highly sensitive to particle size, and it is difficult to impossible to establish test conditions that can yield an absolute value for it.

For nephelometric measurements, fluorimeters are frequently employed. A fluorimeter is a device similar to a standard spectrophotometer except that the photodetector is oriented normal to the direction of the incident radiation (see section on Luminescence). Like turbidimetry, measurements of unknowns are compared to standard calibration curves.

The wave length of the incident light in nephelometry is usually not critical while it is in turbidimetry. This is because one is less concerned with absorption of radiation in nephelometry than in turbidimetry, because turbidimetry is a measurement of the net transmittance of the suspension, and one would like to distinguish between radiation absorption and scattering effects. Thus, while in nephelometry, a white source is frequently used, in turbidimetry, monochromatic sources are selected that exhibit minimum absorption in the test material.

Errors in Measurement

It has been pointed out that the major errors in both turbidimetric and nephelometric analyses derive not from the optical measurements but rather from sample preparation [Kolthoff *et al.* 1969, 993]. This is particularly so because of the particle size sensitivity of the measurements. Thus, theoretical calculations show that suspensions with identical weight concentrations but with particles of radii equal to 0.66, 0.112, and 0.132 μ can exhibit fluctuating relative scattering values of 0.94, 1.76, and 1.59, respectively. Therefore, test conditions must be such that suspension particle sizes are uniform between calibration standards and unknowns.

The factors identified as being critical to achieving consistent particle size include:

1. The concentrations of the ions involved in formation of the suspension
2. The ratio of the concentrations in the solutions mixed
3. The rate and manner of mixing
4. The time between formation of the suspension and the measurement
5. The stability of the suspension, i.e., how rapidly it settles
6. Temperature
7. The presence and nature of other soluble components in the suspension

Applications

The primary environmental applications for turbidimetry or nephelometry are for monitoring outflows from waste treatment plants and for analysis of sulfate in water. The turbidimetric sulfate method is one of a number of sulfate analyses approved by the EPA; details of the procedure may be found in ASTM Standard D 516 and in Section 426C of Standard Methods for Examination of Water and Wastewater. In summary, however, the method consists of precipitating sulfate with barium chloride in a solution containing glycerine and sodium chloride to stabilize the suspension. Measurement is made either turbidimetrically or nephelometrically at a wave length near 400 nm. The minimum detectable concentration of the method is 1 mg/l of sulfate.

LUMINESCENCE AND FLUORESCENCE

Principles

Chemical species may be excited to elevated electronic energy states by heat and chemical reaction as well as by absorption of radiation. The energy transitions that may occur as a consequence are illustrated in Figure 7.16. As can be seen from the illustration, when light, heat, or chemical

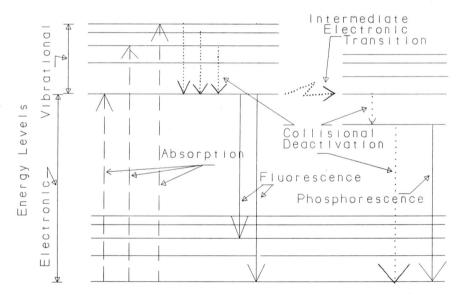

Figure 7.16 Possible energy transitions leading to fluorescence, collisional deactivation, and phosphorescence.

reaction cause excitation to higher electronic states, they also stimulate higher level vibrational modes. In general, the excited species will try to dispose of the excess energy and return to the ground vibrational and electronic states. They may do so by several different mechanisms, but the most common is via energy transfer to neighboring molecules, which is called collisional deactivation. However, when a molecule has an excited electronic state that is relatively stable, part of the elevated energy may dissipate as heat, through relaxation of the higher level vibrational modes, and the remainder may dissipate by emitting light through relaxation of the excited electronic state.

The general name for the light emitting process is luminescence. Luminenscence stimulated by heat is called thermoluminescence and luminescence associated with chemical reactions is called chemiluminescence. Radiation-induced luminescence is divided into two sub-classes called fluorescence and phosphorescence, the distinction between the two being the period during which the emission of light occurs. That is, if light emission is coincident with the stimulating radiation, the phenomenon is called fluorescence; when the emissions persist after the stimulating radiation is removed, the phenomenon is called phosphorescence. The difference in emission times depends upon the paths taken as the excited moieties relax to ground state, with phosphorescence occurring when the energy decay path involves a transition to an intermediate level electronic state.

In our discussion we are concerned primarily with fluorescence and chemiluminescence, since these two phenomena serve as the basis of measurement of sulfur and nitrogen species in the environment.

Fluorescent and Chemiluminescent Intensity

The intensity of radiation-induced fluorescence is directly related to the concentration of the fluorescing species in the medium being tested. This follows from consideration of the absorption of radiation by the fluorescing component. That is, the fraction of light transmitted through a medium containing a fluorescing component is given, according to Beer's law, by

$$\frac{I}{I_0} = \exp(-abC)$$

where I and I_0 are the intensities of the transmitted and incident light, respectively, and a,b, and C are the absorbance, cell path length, and concentration, respectively. The fraction of light absorbed then is

$$1 - \frac{I}{I_0} = 1 - \exp(-abC)$$

and the total amount of light absorbed is

$$I_a = I_0 - I = I_0[1 - \exp(-abC)]$$

But the quantity $[1 - \exp(-abC)]$ may be expanded in a power series

$$[1 - \exp(-abC)] = abC - \frac{(abC)^2}{2!} + \frac{(abC)^3}{3!} - \cdots$$

and, at low concentrations, one may drop the higher order terms. So one may then write

$$I_a = I_0 abC$$

Now the intensity of fluoresence will be proportional to the amount of light absorbed, so

$$I_f = KI_a$$

and

$$I_f = KI_0 abC$$

showing that in dilute solutions of fluorescing compounds, the fluorescent intensity is directly proportional to concentration and to the intensity of the incident radiation.

Chemiluminescence, of course, is derived from chemical reaction rather than radiation absorption. Thus, chemiluminescent intensity is dependent upon the rate at which fluorescing species are chemically generated as well as upon the efficiency with which the energy dissipation processes produce photons. We present, as an example of the factors that may enter the chemiluminescent process, the essentials of an EPA reference method for the determination of nitrogen pollutants in the environment. The method is based on the chemiluminescence of nitrogen dioxide.

Depending upon their oxidation states, the nitrogen pollutants are first converted by either combustion or catalytic reduction to nitric oxide and then reacted with ozone to form excited nitrogen dioxide

$$NO + O_3 \overset{k_1}{=} NO_2^* + O_2 \qquad (7.10)$$

where k is a reaction rate constant defined by the rate at which the reaction products are produced or reactants disappear (see below).

But not all of the nitric oxide that reacts with ozone produces excited nitrogen dioxide, so there is a competing reaction analogous to eq. 7.10

$$NO + O_3 \overset{k_2}{=} NO_2 + O_2 \qquad (7.11)$$

which produces non-excited nitrogen dioxide at a different rate than reaction 7.10.

In addition, although each of the excited nitrogen dioxide molecules is a potential photon source, not all of them produced in reaction 7.10 necessarily relax via that path. Some may return to ground state via collisional deactivation. Thus, there are two competing reactions that represent the relaxation of the excited nitrogen dioxide molecules

$$NO_2^* \overset{k_3}{=} NO_2 + photon \qquad (7.12)$$

and

$$NO_2^* + M \overset{k_4}{=} NO_2 + M \qquad (7.13)$$

where M represents the concentration of components to which the excited nitrogen dioxide may transfer its energy without emitting any photons.

Basic reaction kinetic considerations teach that the rate at which the nitric oxide in eq. 7.10 is converted to excited nitrogen dioxide is proportional to the concentrations of both the nitric oxide and ozone in the reaction medium. That is, the rate of decrease of nitric oxide in the reaction medium is given by

$$\frac{d[NO]}{dt} = -k_1[O_3][NO] \qquad (7.14)$$

where the square brackets denote concentration.

But in the analytical procedure, the concentration of ozone in the reaction medium is maintained constant so that on integrating eq. 7.14, one gets

$$log(f_t) = -k_1[O_3]t$$

or

$$f_t = exp(-k_1[O_3]t)$$

where f_t represents the fraction of the initial concentration of nitric oxide that reacts, i.e., $[NO]_t/[NO]_{initial}$, after the time t. The fraction of the total nitrogen dioxide that may be produced by the reaction, therefore, is given by

$$F_t = 1 - \exp(-k_1[O_3]t) \qquad (7.15)$$

However, to estimate the photon producing capability of the excited nitrogen dioxide generated, eq. 7.15 must be modified by the competing reactions.

Using the same reaction kinetic reasoning as in eq. 7.14, one can also write rate equations for the production of excited and non-excited nitrogen dioxide. That is,

$$\frac{d[NO_2^*]}{dt} = k_1[NO][O_3]$$

and

$$\frac{d[NO_2]}{dt} = k_2[NO][O_3]$$

And the fraction of the total amount of nitrogen dioxide produced by nitric oxide/ozone reaction that is excited is, therefore

$$\frac{k_1}{k_1 + k_2}$$

Similarly, using eqs. 7.12 and 7.13, one can write for the rates at which the excited nitrogen dioxide molecules relax to ground state

$$\frac{d[NO_2^*]}{dt} = k_3[NO_2] \qquad \text{(with photon emission)}$$

and

$$\frac{d[NO_2^*]}{dt} = k_3[NO_2] \qquad \text{(with photon emission)}$$

Therefore, the fraction of excited molecules that emit photons is given by

$$\frac{k_3}{k_3 + k_4M}$$

The net photon generation rate for this overall chemiluminescent reaction, then, is given by

$$\Phi = \{1 - \exp(-k_1[O_3]t)\}\left\{\frac{k_1}{k_1 + k_2}\right\}\left\{\frac{k_3}{k_3 + k_4M}\right\} \qquad (7.16)$$

An interesting consequence of eq. 7.16 is that it suggests that improved photon generation can be obtained by reducing the overall pressure of the system being observed. That is, by reducing [M], one can increase the probability that the excited NO_2 molecules will relax by emitting photons rather than by transferring their excess energy to M. In fact, measuring systems operating at reduced pressure are capable of determining nitrogen pollutants at the sub-ppm level.

Instrumentation

A schematic of a fluorimeter is shown in Figure 7.17. It can be seen to be quite similar to an absorption spectrophotometer except for the position of the detector, which is oriented at an angle of 90° to the axis of the stimulating radiation. In addition, fluorimeters require two filter or light dispersion devices (monochromators), one for the source and one for the detector. This is required for two reasons. First, the wave length of the fluorescent radiation is always different and lower than that of the source radiation and second, the detector monochromator filters out any source radiation that may be scattered by the sample into the detector light path. Like absorption spectrophotometric determinations, fluorometric measurements are alway referenced to standard calibration curves.

In contrast to the fluorimeter, measurement of chemiluminescence requires only a detector, usually a photomultiplier. But the instrumentation becomes somewhat more complex due to the requirement to control the light-stimulating chemical reaction. A typical instrument arrangement for the chemiluminescent determination of nitrogen compounds is illustrated in Figure 7.18.

Nitrogen compounds from the sample are passed first into a reaction chamber where they are converted to nitric oxide. The nitric oxide produced is then swept into a second reaction chamber through which a controlled flow of ozone is maintained and in which resides the detector window of a photomultiplier. Light output of wave length greater then 600 nm, integrated over time, is directly related to the amount of nitric

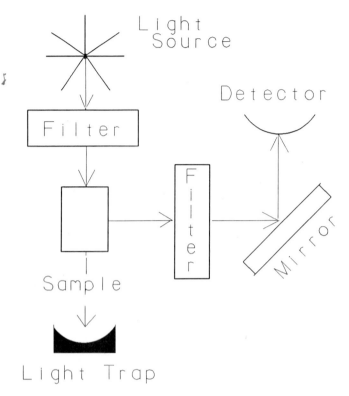

Figure 7.17   A simplified schematic of a fluorimeter.

oxide that enters the second reaction chamber. However, as in fluorime-try, measurements are always referred to standard calibration curves.

Sensitivity and Specificity

Both the sensitivity and specificity of luminescent measurements are higher than that of standard absorption spectrophotometric methods. Sensitivity is improved because while one is measuring reductions in large background light signals in absorption spectrophotometry, i.e., measuring relatively small differences between large background signals, one is measuring light signals against effectively black backgrounds in luminescence. The sensitivity of luminescent measurements is said to be as much as 10 to 10^3 better than light absorption methods [Olsen 1975, 404].

Specificity is improved simply because of the fact of fewer materials that exhibit luminescence. Furthermore, fluorometric measurements, in

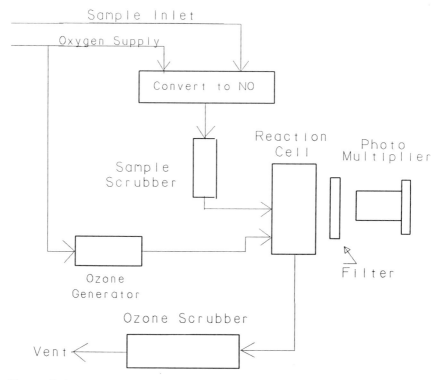

Figure 7.18   Block diagram showing a typical instrument layout for the chemiluminescent measurement of nitrogen compounds.

contrast to light absorption methods, require selection of separate excitation and detection frequencies. Therefore, at least in principle, if the luminescent spectra of two materials differed in either parameter, one could measure one in the presence of the other.

Applications

Chemiluminescent measurement of nitrogen pollutants has already been discussed. Details of the procedures required by EPA in applying the technique to measurement of atmospheric nitrogen dioxide are presented in the Code of Federal Regulations [40 CFR Part 50.12, Appendix F].

   In environmental applications, fluorescence is an approved reference method for the determination of sulfur dioxide in the atmosphere [Quality Assurance Handbook, Vol. II, Section 2.9], and the technique has been applied to determination of elemental sulfur in liquids and solids. Measurement is based on the excitation of SO_2 by ultraviolet radiation, which produces a product that fluoresces in the nearer ultraviolet region. For the analysis of solids and liquids, samples are burned in oxygen to form carbon dioxide and sulfur dioxide, and the product gases are passed into the fluorimeter.* The sensitivity of fluorescent sulfur determinations is of the order of a few ppm.

X-Ray Fluorescence

We mention briefly x-ray fluorescence as an alternative — though sensitivity–limited, method to determine elemental sulfur in liquids and solids. In x-ray fluorescence, samples are subjected to incident x-radiation, which promotes electrons out of the inner orbitals of the atoms. Relaxation then gives rise to emission of secondary x-rays with energies characteristic of the exited element. The secondary x-rays are measured with a device called a goniometer which, like prisms or gratings for the optical region of the spectrum, disperses x-radiation into its components. The dispersed rays may be detected by Geiger counters, scintillation detectors, and similar particle counting devices. The sensitivity of x-ray fluorescence instruments is in the low percent region. The specificity of the method, however, is excellent because of the distinct energy characteristics of the emissions from different elements.

MASS SPECTROMETRY

Principles

In the earlier section on Fourier Transform Infrared Spectroscopy, mention was made of so-called hyphenated analytical techniques. The combination of mass spectrometry with gas chromatograpy (GCMS) is another such technique: it is also a technique that is approved by the EPA for the analysis of a large number of pesticides and other organic compounds in water [40 CFR, Part 136.3]. We present here a brief discussion of the

---

*A caution about the use of combustion methods to convert sulfur compounds to sulfur dioxide is worth mentioning. That is, combustion of sulfur compounds does not necessarily quantitatively convert sulfur to sulfur dioxide, and the conversion efficiency is dependent upon the chemical nature of the sample. Thus, one cannot use standard gas mixtures for calibration. Rather, one must prepare calibration curves from combustion of samples with overall chemistry similar to that of the unknowns.

principles of mass spectrometry in anticipation of discussion of its use in combination with gas chromatography in a later section.

In contrast to the spectroscopic methods that we discussed earlier, in which analysis is based on the dispersion of electromagnetic radiation into its component frequencies, in mass spectrometry, analysis is based on the dispersion of the material under test into its mass components. Thus, for example, the mass spectrogram of a naturally occurring gas like oxygen includes a spectral line for each of the naturally occurring isotopes of oxygen; and for naturally occuring hydrogen, one obtains a spectrum that includes lines corresponding to the mass numbers of hydrogen, deuterium, and tritium.

The separation of species by mass in a mass spectrometer has its foundations in the fact that an electrically charged particle moving in a straight line that passes between the poles of a magnet is deflected by the magnetic force and traverses a circular trajectory while under the influence of the magnetic field. The radius of the circular trajectory is related to the magnetic force imposed upon the charged particle as well as to its charge and kinetic energy, i.e., its mass and velocity, as it enters the magnetic field. Thus, if charged particles of varying mass but of equal kinetic energy enter the influence of a magnetic field, they will traverse circular trajectories with radii that depend upon their mass and charge and they will separate accordingly.

The process by which mass separation is accomplished is shown schematically in Figure 7.19. The sample is vaporized and the vapor particles are introduced into a high vacuum chamber where they are first bombarded with high energy electrons* or allowed to react with other gaseous ions.† As a result, the sample molecules are ionized. The ions produced are then passed into the analyzer section of the spectrometer where they are all accelerated by a high potential electric field and acquire kinetic energy given by

$$\frac{mv^2}{2} = eV$$

where e is their charge, v their velocity, m their mass, and V is the strength of the electric field. Thus, their velocity upon exiting the electric field is

---

*Called electron impact mass spectrometry.
†Called chemical ionization mass spectrometry.

given by

$$v = \left(\frac{2eV}{m}\right)^{1/2} \tag{7.17}$$

The ions are then passed into a magnetic field oriented perpendicular to their direction of motion, where they are subjected to a force which, according to Ampere's law, is given by Hev, where H is the magnetic field strength. As a result, they are deflected into circular trajectories of radius (r) with a centrifugal force of (mv^2/r), which must be equal to the magnetic force that gave rise to the path deflection. That is

$$\frac{mv^2}{r} = Hev \tag{7.18}$$

The dependence of their path radii on mass, charge, and magnetic field strength can then be seen by substituting the value of v from eq. 7.17 into eq. 7.18 and rearranging

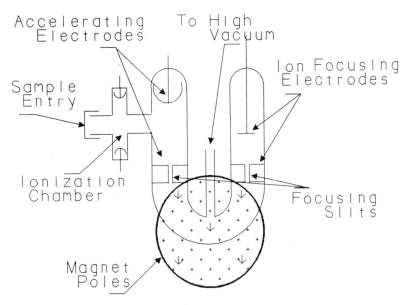

Figure 7.19   Schematic of a Dempster type mass spectrometer.

$$r = \frac{mv}{He} = \frac{1}{H}\left(\frac{2Vm}{e}\right)^{1/2}$$

And, of course, it is the path dependence upon the m/e ratios and field strength that serves as the basis of separation. By varying the magnetic field strength, it is possible to successively focus ions of varying m/e on the spectrometer's detector. Now, although the ionization process may produce multiply charged ions, the geometry and magnetic field strength of the instrument are arranged so that ions with charges greater than unity are never focused on the detector. Thus, the signals arriving at the detector will be related to particle mass only, and their intensity will be proportional to the concentrations of the charged species produced in the ionization process.

Applications

The primary environmental application for mass spectrometry is for identification of organic contaminants in water. Unfortunately, the method by itself is not really appropriate for the analysis of mixtures of organics that one might normally find in contaminated waters. This is because of the complexity of the mass spectra that even pure and isolated organic compounds exhibit, because the ionization process produces a host of ion products besides the ionic counterparts of the parent compounds. That is, in the ionization process, some of the molecules of the parent compounds may fragment into elemental and small molecule ions; and the fragments can react with themselves, other fragments, and the parent ions to produce still other ions. Additionally, the unstable configuration of the parent ions can lead to molecular rearrangements and the formation of still other ionic moieties. Consequentially, the mass spectra of organic compounds tend to be complex, and a mixture of organics produces spectra with still greater complexity.

A fortunate fact of nature, however, is that the spectra obtained for single compounds are generally unique to those compounds. That is, no two different organic compounds display identical spectra, although closely related molecules may exhibit similar fragmentation patterns. Because of the uniqueness of the spectra, identification of compounds can be made by matching observed spectra with the spectra of known materials. Large compilations of mass spectra for organic compounds have been assembled, and computer search techniques are available to search among them to find matches. One caution that must be mentioned, however, is that the mass spectral patterns are dependent upon the test conditions established for each spectrometer. So, in using matching techniques to identify compounds, one must be certain that the reference spectra and the test spectra were obtained under similar conditions.

Because of the complexity of the spectra obtained for mixtures of organics, identification of individual components is essentially impossible. However, if the identity of the components is known, the spectra may be used to carry out quantitative analysis. This is because the spectra of individual compounds are, in general, independent of the chemical environment in which they reside. Thus, the observed spectrum of a mixture is the sum of the individual spectra, i.e., the individual line intensities are the sum of the individual contributions of each of the mixture's components. The procedure used to do quantitative analysis is as follows.

If the signals received at the spectrometer's detector are I_i, I_j, I_k, etc., for mass numbers i, j, and k, etc., and x, y, and z, etc., represent the compounds in the mixture, then one can write a series of simultaneous equations of the kind

$$I_i = i_{ix}P_x + i_{iy}P_y + i_{iz}P_z + \cdots$$

$$I_j = i_{jx}P_x + i_{jy}P_y + i_{jz}P_z + \cdots$$

$$I_k = i_{kx}P_x + i_{ky}P_y + i_{kz}P_z + \cdots$$

where the P_i are the partial pressures of each of the mixture's components in the sample, and the i_n are the signals received at mass numbers i, j, and k, etc., in taking spectra of the pure compounds x, y, and z at known sample pressures. Solution of the equations yields the values of P_i and the concentrations of the mixture's components.

However, rarely can the enviromental analyst count on knowing the composition of his contaminated water sample. So, mass spectrometry has little utility for him independent of supplementary identification procedures, and it is to that end that it is combined with gas chromotography to provide a powerful environmental analytical tool.

8

Electrochemical Methods

INTRODUCTION

Like spectroscopy, electrochemical methods consititute a generic class of analytical procedures, not any one specific technique. And like spectroscopy, their breadth of applications is vast, covering general quantitative chemical analysis as well as being integral to a number of environmental analytical procedures. Unlike spectroscopy, however, electochemical methods can not generally be used to identify materials. That is, electrochemical methods are nonspecific. This is because electrochemical properties are generally more dependent upon the number of molecular or ionic entities present in an analysate rather than upon their chemistry. One technique, however, polarography, can be used to profile a mixture of oxidizable compounds according to their electrochemical decomposition characteristics. But the primary strength of electrochemistry is for the quantitative analysis of isolated compounds or mixtures of known composition.

Electrochemical methods are generally based on measurement or control of electrical potential or electrical conductivity. For the former, one is concerned with electron transfer processes resulting from oxidation/reduction reactions occurring at the interface of a solution and a conductor. For the latter, one is concerned with the electric current associated with the transfer of charge resulting from the transport of ionic species through a solution.

In our discussion, we will consider three different kinds of electrochemical methods: those in which electrical potential is directly related to analyte concentration, those in which electrical conductivity is directly related to analyte concentration, and kinetic methods in which an electrical potential is applied to an analysate and the resulting current or current-time response is related to analyte concentration.

POTENTIOMETRY

Principles of Potentiometric Cells

Electrical potentials are generated in electrochemical cells as a result of oxidation/reduction reactions. Recall that oxidation and reduction are defined as the loss and gain of electrons, respectively, by a chemical entity. When one substance is reduced, however, another must be oxidized so as to provide the electrons for the reduction process. Formally, the substance being reduced is called the oxidizing agent, or the oxidant, and the one being oxidized is called the reducing agent, or the reductant.

Oxidation/reduction or redox reactions, as applied in electrochemical analysis, are carried out in cells which consist, essentially, of two conducting electrodes immersed either in a single solution or in two different solutions that are in electrical contact with one another. Depending upon the chemical nature of the electrodes and the consituents of the solution or solutions, the redox reactions in the cell may occur spontaneously. As such, the cell may serve as a source of electrochemical energy and is called a galvanic cell. Contrariwise, it may be necessary to apply electrical energy to the cell to induce the redox reactions; such a cell is called electrolytic. A battery is an example of a galvanic cell while an electroplating process is an example of an electrolytic cell.

A typical potentiometric cell is shown in Figure 8.1. The cell contains a solution of hydrochloric acid. One electrode is made of a platinum wire that is mounted in a glass tube through which hydrogen gas, at a pressure of one atmosphere, is passed and allowed to bubble through the HCl solution so that it is in intimate contact with the platinum wire and the solution. The second electrode is a silver wire upon which has been overlayed a coating of silver chloride. Each of the electrodes is said to constitute a *half-cell*, and the combination of the two half-cells constitutes the total cell. This is a galvanic cell, and a voltmeter connected between its two electrodes exhibits a voltage that is related to the concentration of HCl in the solution. That relationship is a consequence of the redox reactions that can occur in the cell.

The redox reactions that do occur do so in this cell because hydrogen gas in the presence of platinum is a stronger reducing agent than silver chloride. Given the opportunity, therefore, it will reduce the silver chlor-

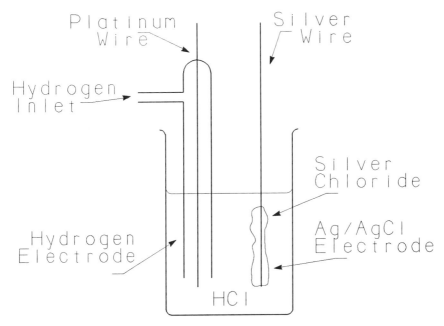

Figure 8.1   A simple electrochemical cell containing hydrogen and a silversilver chloride electrodes.

ide to metallic silver spontaneously. So, if instead of measuring the potential between the two electrodes, we short circuited them, current would flow through the cell, with the source of the current being the electrons produced by the hydrogen gas as it was oxidized to hydrogen ions. The electrons generated would then flow through the external circuit to the silver chloride, which would be reduced to metallic silver.*

The electrochemist's conventional representation of the cell in Figure 8.1 is written

$$Pt/H_2(p = 1\ atm)/HCl(m = C)/AgCl/Ag \qquad (8.1)$$

with the single slashes representing the interfaces between the cell components.

---

*Silver chloride is a crystalline solid composed of silver (Ag^+) and chloride (Cl^-) ions. The Ag^+ entities in it are reduced to metallic silver while the chloride ions are discharged to the solution. The electric charge of the discharged Cl^- is neutralized by the H^+ ions generated by oxidizing the hydrogen gas.

The reactions underway at the electrodes while the current flows are

$$\tfrac{1}{2}H_2 = H^+ + e^-$$

and

$$AgCl + e^- = Ag + Cl^-$$

and the net reaction for the cell is

$$\tfrac{1}{2}H_2 + AgCl = H^+ + Cl^- + Ag$$

noting that the conventional representation of an electrochemical cell always shows the oxidizing process at the left and the reduction process at the right. Also note that the oxidation of a formula weight of silver produces the same number of electrons as the oxidation of half a formula weight of hydrogen gas. This relationship serves as the basis of the definition of the chemist's *equivalent weight*. That is, an equivalent weight (or an equivalent) of any chemical compound is that amount that will yield, when it is oxidized, enough electrons to produce one half a formula weight of hydrogen gas from hydrogen ions.* That number of electrons, incidentally, is also called an equivalent of electrons and is equal to a "Faraday" of electricity or 96,494 coulombs of electric charge.

The relationship between the reacting species in an electrochemical cell and its electrical potential becomes accessible by consideration of the work that the cell can do if current flows through it.[†] By definition, if we allow enough current to flow through Cell (8.1) to make an equivalent of HCl (as H^+ and Cl^- ions), an equivalent of electricity passes. The work done, therefore, is given by

$$w = EIt = EF$$

where E is the voltage between the two electrodes, I is the current, t the time for an equivalent to be produced, and F the Faraday.[‡]

---

*An equivalent of the metal of a divalent ion, for example, would be equal to half its atomic weight and that of a trivalent ion, one third of its atomic weight

[†]It is important to note that real measurements of potentiometric cells are made under conditions of zero current so as to inhibit reactions that may change the chemistry of the cell contents. Consideration of current flow is simply a device used to derive the relationships governing cell behavior.

[‡]Current is defined as charge per unit time. So the product (It) is the amount of charge passed and is equal to an equivalent of electricity or the Faraday.

Now, from chemical thermodynamics, the work done by the cell is also represented by the negative of the *free energy change* that occurs in the production of the HCl, and the free energy change is given by the difference in free energy between the products and the reactants of the reaction. That is

$$-\delta G_{total} = -(\delta G_{prod} - \delta G_{react}) = EF*$$

It can also be shown that the free energy is related to the concentration of the cell constituents through

$$\delta G_{total} = \{\delta G_{0_{prod}} + RT \log[Cl^-][Ag][H^+]$$
$$- \{\delta G_{0_{react}} + RT \log[AgCl][H_2]^{1/2}\} = (-EF) \qquad (8.2)^\dagger$$

where δG_0 in each of the terms in braces represents the standard free energies of the products and reactants (see footnote "*" below) and R and T are the universal gas constant and the temperature, respectively.

Combining the log terms in eq. 8.2 and dividing by F, therefore, yields the cell potential as a function of cell constituent concentrations

$$E = E_0 - \frac{RT}{F} \log \frac{[Cl^-][H^+][Ag]}{[H_2]^{1/2}[AgCl]} \qquad (8.3)$$

---

*The deltas are used because there is no absolute free energy scale. Free energy is always measured from the so-called standard states of the elements, which are assigned a value of zero in their most stable configurations and at 25° and 1 Atm. The standard free energy of a compound is the free energy change that occurs in its formation from its elements. More complete explanations may be found in any introductory text in physical chemistry or chemical thermodynamics [Moore 1950; Klotz 1950].

†The brackets around the chemical symbols in this and all the following equations denote the concentrations of the components. However, strictly speaking, one should use the thermodynamic activity rather than concentration. Thermodynamic activity is a term that represents the real, in contrast to the ideal, behavior of solutes, when the property being affected by them is dependent upon their number in solution rather than upon their chemical nature. That is, X dissolved particles generally do not behave as if there were really X of them there, because dissolved molecular and ionic particles interact with one another except at very high dilutions. Thus, for purposes of exact calculations, solution concentrations must be modified by multiplying them by a dimensionless, experimentally determined factor called the activity coefficient to reflect their non-ideal behavior. The product of the activity coefficient and the concentration is called the activity. To simplify our discussion, however, we shall assume that concentration and activity may be freely substituted for one another.

where $E_0 = 1/F(\delta G_{0_{prod}} - \delta G_{0_{react}})$ is known as the standard potential for the cell reaction.

Now, since the only processes underway in Cell (8.1) are the two electrode reactions, one can split eq. 8.3 into two parts, each reflecting only one of them. Thus, one can write

$$E = E_{0_H} - \frac{RT}{F} \log \frac{[H^+]}{[H_2]^{1/2}} + E_{0_{AgCl}} - \frac{RT}{F} \log \frac{[Ag][Cl^-]}{[AgCl]} \qquad (8.4)$$

where the first two terms may be identified with the hydrogen electrode and the second two with the silver/silver chloride electrode. That is, each of the terms in eq. 8.4 may be identified with the separate electrodes, with each of them exhibiting a single electrode or *half-cell potential* that is dependent upon the concentrations (or activities) of its individual redox reactants. The E_0 are called the standard potentials of the half-cell reactions.

Reference and Reversible Electrodes

With some reflection it becomes evident, however, that single electrode potentials have no meaning independent of a partner electrode system, because electrical measurements can only be made of differences of potential between electrodes. To accomodate the problem, the hydrogen half cell reaction has been selected as a reference system. That is, its standard potential (E_{0_H}) has been defined to be equal to zero. Then, when the concentration of the H^+ ions (or HCl) is made equal to one molar (activity = 1) and the hydrogen gas pressure is fixed at one atmosphere (activity = 1) in its cell reaction, its total potential also becomes zero, as reference to eq. 8.4 illustrates. That specific configuration, which is called the *Standard Hydrogen Electrode (SHE)*, is the configuration to which all other half-cell reactions are referred.

With that standard defined, eq. 8.4 reduces to

$$E = E_{0_{AgCl}} - \frac{RT}{F} \log \frac{[Ag][Cl^-]}{[AgCl]} = E_{0_{AgCl}} + \frac{RT}{F} \log \frac{[AgCl]}{[Ag][Cl^-]}$$

where the expression on the far right is called the *Nernst Potential* for the Ag/AgCl half-cell. A generalized expression for the Nernst potential for any redox couple is given by

$$E = E_0 + \frac{RT}{nF} \log \frac{[\text{oxidized component}]}{[\text{reduced component}]} \qquad (8.5)$$

where n refers to the number of electrons involved in the oxidation reaction.

Clearly, although the hydrogen electrode is the defined standard, the silver-silver chloride half-cell might also serve as a reference electrode. In fact, in general, any "reversible couple" may serve as a reference electrode where, by a reversible couple, we mean an electrode system that simply reverses its redox reaction direction, maintaining the identical reactants and products regardless of the direction, as increments of current are passed back and forth across it.

There are several classes of reversible electrodes. A metal dipping into a solution of one of its salts is one type, with the metal being reversible to its ions. A second type is that of the SHE, where an inert metal like platinum serves as a conductor for electrons generated by the oxidation or reduction of a gas. Some oxygen and chlorine electrodes fall into this category. An inert metal in contact with a solution of ions containing two valence states also constitutes a reversible electrode. Thus, for example, a platinum conductor immersed in a solution of ferric and ferrous chloride is said to be a reversible Fe^{++}/Fe^{+++} electrode. And finally, there is the type in which a metal in contact with one of its insoluble salts is immersed in a solution containing the anions of the insoluble salt. Ag/AgCl and calomel (see section on Liquid Junction Potentials) electrodes fall into the last class as well as electrodes reversible to the other halide ions and sulfate.

Potentiometric Analysis

Combinations of pairs of half-cells, as shown in Figure 8.2, constitute analytical potentiometric cells, with one half-cell containing a reversible electrode serving as a reference and the other half-cell containing a reversible electrode serving as an indicator. Typically, a liquid junction, in the form of a porous membrane or a tube containing a gelled, concentrated solution of an electrolyte (a salt bridge), joins the two half-cells and serves as their electrical connection as well as a device that prevents the internal solutions of the two electrode systems from mixing. In that way, changes in the concentration of the internal solution of the indicator electrode, i.e., in the test chamber, do not affect the half-cell potential of the reference. However, in analytical cells of this configuration, the liquid junction also contributes to the cell potential. Thus, the total potential is given by

$$E = E_{ref} - E_{test} + E_j \qquad (8.6)$$

where E_j is called the liquid junction potential.

In practice, cells are arranged so that E_{ref} is a constant and E_j is either constant or a very small value. So, the potential of an analytical cell, in

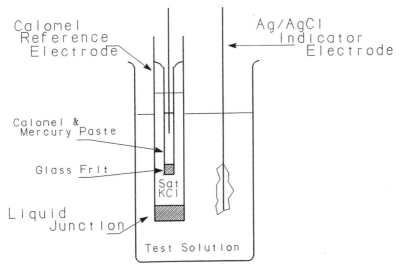

Figure 8.2   An electrochemical cell with a liquid junction — in this case the junction can be a porous glass frit or a porous membrane.

general, is given by

$$E = K - \left(E_{0\text{test}} + \frac{RT}{nF} \log \frac{[\text{oxidized component}]_{\text{test}}}{[\text{reduced component}]_{\text{test}}} \right)$$

i.e., variations in its potential are only a function of the concentrations of the electroactive components in the test chamber.

Liquid Junction Potentials

The cell shown in Figure 8.2 may be represented by the cell diagram

$$\text{Hg/Hg}_2\text{Cl}_2\text{/KCl(sat)//KCl(m = b)/AgCl/Ag} \qquad (8.7)$$

where the double slash denotes the liquid junction. The reference electrode in this case is the very commonly used saturated calomel electrode (SCE),* and the Ag/AgCl electrode serves as the indicator element, re-

---

*The saturated calomel electrode is composed of mercury and mercurous chloride (calomel) in contact with a saturated solution of potassium chloride. Its half-cell reaction is $\text{Hg} + \text{Cl}^-(\text{sat}) = 1/2\,\text{Hg}_2\text{Cl}_2 + \text{e}^-$. Its half-cell potential is given by $E = E_0 - (RT/F)\log(1/[\text{KCl}]_{\text{sat}})$.

versible to the chloride ion of the KCl in the test solution. This cell may be used to illustrate the relationship of the liquid junction potential to the total cell potential.

We determine the potential of Cell (8.7) in the same fashion as before by considering the processes that it undergoes when current passes through it. That is, the mercury is oxidized to calomel and the AgCl is reduced to metallic silver. The result in this cell, however, is that the chloride ions consumed in the formation of the calomel leave excess potassium ions, i.e., excess positive charge, in the reference electrode chamber, and the reduction of the AgCl adds excess negative charge in the form of chloride ions to the right chamber. But it can be shown that charge separation between ions in solution is, for all practical purposes, impossible [Guggenheim 1950, 331], so potassium ions migrate to the right and chloride ions to the left across the liquid junction to achieve electrical neutrality. This, of course, is the mechanism by which charge is transferred through the interior of the cell to complete the electrical circuit: it also represents a work term in the cell process associated with the liquid junction. Thus, following eq. 8.6, the total cell potential for 8.7 which, incidentally, is referred to as a cell with transference, is given by

$$E = E_{Hg} - E_{Ag} + E_j \qquad (8.8)$$

where E_{Hg} and E_{Ag} are the electrode potentials of the calomel and silver chloride half-cells, respectively.

If a full equivalent of electricity were to pass through Cell (8.7), an equivalent of excess potassium ions would be generated in the left chamber, an equivalent of excess chloride ions would be produced in the right, and the ions would co-diffuse across the liquid junction. Ions, however, move independently of one another (see section on conductance below), and they diffuse to and through the junction at rates that are dependent upon their ionic mobilities.* Each of them, therefore, carries only that fraction of the cell current that their mobility permits. That fraction is called the ionic transference number and is conventionally denoted t_+ for the positive species and t_- for the negative species. Clearly $(t_+ + t_-) = 1$ and, in general, for a system with many different kinds of ions present $\Sigma t_i = 1$. So, in Cell (8.7), t_+ equivalents of potassium ions diffuse from the left chamber to the right while t_- equivalents of chloride move from the right chamber to the left. The t_- equivalents of chloride ions neutralize the remaining t_- equivalents of potassium in the left

---

*Ionic mobility is defined as an ion's velocity in a unit electric field. It is dependent upon ionic radius and charge.

chamber, and the remaining t_+ chloride ions in the right chamber neutralize the t_+ equivalents of incoming potassium ions. The net result of the current flow in Cell (8.7) then, if one assumes the chambers to be so large that the cell processes do not change their initial concentrations, is the transfer of t_+ equivalents of potassium and chloride ions from a saturated KCl solution to one of concentration $m = b$.

The transfer of a solute from a solution of one concentration to another requires the expenditure of work, and the free energy change associated with it is given by

$$\delta G = \delta G_0 + t_+ \{ \delta G_{K^+_{sat}} - \delta G_{K^+_{m=b}} + \delta G_{Cl^-_{sat}} - \delta G_{Cl^-_{m=b}} \} \qquad (8.9)$$

where δG_0 is the standard free energy difference between the calomel and Ag/AgCl half-cells. And since $\delta G = -EF$, by analogy with eq. 8.2, we get

$$E = E_0 - \frac{t_+ RT}{F} \log \frac{[K^+]_b [Cl^-]_b}{[K^+]_{sat} [Cl^-]_{sat}}$$

and

$$E = E_0 - \frac{2t_+ RT}{F} \log \frac{[KCl]_b}{[KCl]_{sat}} \qquad (8.10)$$

where, since $[K^+] = [Cl^-] = [KCl]$, we have taken $[K^+][Cl^-]$ to be equal to $[KCl]^2$.

Equation 8.10 gives the total potential related to the work the cell does generating and moving the KCl from one chamber to the other across the liquid junction. To separate the work represented only by the movement across the liquid junction from the total potential, one must do some simple thermodynamic reasoning. That is, one must hypothesize a process in which the same expenditure of energy used in Cell (8.7) to transfer KCl across the liquid junction is used to transfer a full eqivalent of KCl from the left chamber to the right without passing through the liquid junction.* The free energy change associated with that kind of transfer would be given by

$$\delta G_{hyp} = \delta G_0 + RT \log\{[KCl]_b / [KCl]_{sat}\} \qquad (8.11)$$

---

*For example, it might be vaporized from one chamber and condensed into the other.

But a basic premise of thermodynamics is that a change in free energy accompanying a change in state (in this case moving the KCl from a solution of one concentration to another) is independent of the path taken. So δG_{hyp} in eq. 8.11 may also be set equal to $-E_{hyp}F$, and the hypothetical process can be represented by an electrical potential given by

$$E_{hyp} = E_0 - \frac{RT}{F} \log\{[KCl]_b/[KCl]_{sat}\} = E_{Hg} - E_{Ag}$$

Then, from eqs. 8.8 and 8.10

$$E_j = E_0 - \frac{2t_+RT}{F} \log\frac{[KCl]_b}{[KCl]_{sat}} - E_0 + (RT/F) \log\frac{[KCl]_b}{[KCl]_{sat}}$$

and we find the potential for the liquid junction between the reference and test of Cell (8.7) to be given by

$$E_j = (1 - 2t_+) \frac{RT}{F} \log\frac{[KCl]_b}{[KCl]_{sat}} = (t_- - t_+) \frac{RT}{F} \log\frac{[KCl]_b}{[KCl]_{sat}} \quad (8.12)$$

In fact, eq. 8.12 is a general form for any liquid junction between identical univalent electrolytes. It also suggests how the liquid junction potential may be minimized. That is, one may use a highly concentrated electrolyte solution in the junction, choosing one whose ions have similar mobilities. In that way, the bridge electrolyte is made to dominate the current flow through the junction, and the pre-logarithmic term $(t_- - t_+)$ tends toward zero. It is for this reason that concentrated solutions of potassium chloride or potassium nitrate are frequently used in salt bridges, because the transference numbers* of Cl^-, NO_3^-, and K^+ are nearly equal [Robinson and Stokes 1959, 465].

Ion Selective Electrodes

An additional and important consequence of eq. 8.12 is that it illustrates that the liquid junction potential is independent of E_0 and, thus, independent of the cell's specific electrode reactions. Yet, it is dependent upon the concentrations of the solutions with which the junction interfaces, suggesting that E_j, itself, can serve as an analytical probe. In fact, it is that property of junction potentials that is the basis of operation of a class

---

*Actually the equivalent conductivity, see Conductometry section.

of analytical devices called ion selective electrodes, which may be used to measure the concentrations of selected ionic components in a mixture of electrolytes.

An ion selective electrode is, essentially, a reversible electrode which, instead of interfacing to an unknown analyte solution through a liquid junction, does so through a membrane that permits the transfer of only a selected ionic moiety. A cell diagram for a typical ion selective configuration is

$$Ag/AgCl/KCl(sat)\backslash\backslash$$

where we have used the double backslash to denote a membrane selective to potassium ion.

To understand the operation of such an electrode system, however, it is necessary to first develop a more general equation for the liquid junction potential. Thus, consider a liquid junction separating two solutions of overall composition C and C + dC that contain an arbitrary mixture of electrolytes. The transference numbers of the ions in the solutions are t_i, and z_i is their charge. Therefore, passage of an equivalent of electricity through the junction is accompanied by the transfer of t_i/z_i equivalents of each of the ionic species present. By analogy with eq. 8.9, the free energy change associated with the transfer across the junction is given by the sum of the free energy changes associated with the transfer of each of the (i) ions from a region of concentration C_i to one of $(C_i + dC_i)$. Thus, one can write

$$dG = RT\Sigma(t_i/z_i)d \log C_i$$

Equating dG to $-Fd(E_j)$ then gives

$$dE_j = -\frac{RT}{F}\sum \frac{t_i}{z_i} d \log C_i \qquad (8.13)$$

which can be seen to be a more general case of eq. 8.12.

But from its definition, in an ion selective membrane, all the t_i are zero except for the selected ionic moiety, the transference number of which is unity. That is, the junction potential for an ion selective membrane is given by

$$dE_{jsel} = \frac{RT}{F} d \log C_{sel}$$

which, upon integration is

$$E_{jsel} = \frac{RT}{F} \log \frac{[C]_a}{[C]_b}$$

where a and b refer to the bounding solutions.

Now if one used a potassium selective electrode in the cell

$$Ag/AgCl/KCl(sat)\backslash\backslash KCl(C = ?),M_iCl(C_i = ?)//KCl(sat)/Hg_2Cl_2/Hg$$

where M_i refers to any other ionic moieties besides potassium present, its potential would be

$$E = E_{AgCl} - E_{Hg} + E_j + E_{jsel}$$

That is, it would be given by the difference in the half-cell potentials of the reversible electrodes plus the contributions from the liquid junction and the ion exchange membrane. Since neither of the reversible electrodes suffers any change in state when current flows, the difference in potential between them is just the difference in their standard potentials. And, as discussed above, the value of E_j may be minimized. So, the net potential of the cell is a function only of the difference in the concentration of the potassium ions in the unknown solution and the internal solution of the ion selective electrode. That is

$$E = K - \frac{RT}{F} \log [K^+]_?/[K^+]_{sat}$$

Environmental Applications of Potentiometric Cells

Probably the most familiar application of potentiometric cells is for the measurement of pH. Since pH is defined as the negative of the logarithm of the hydrogen ion concentraton, i.e., $pH = -\log[H^+]$, its value is directly accessible from the measurement of the potential of a hydrogen half cell, and it is clear that the hydrogen electrode described earlier could be used. But the hydrogen electrode is an inconvenient device to deal with, so pH is most commonly measured with a *glass membrane* electrode that behaves as ion selective device selective to hydrogen ions. Figure 8.3a is a diagram of the glass pH electrode that is represented by the cell schematic

$$Ag/AgCl/HCl(0.1m)\backslash\backslash$$

Any reference electrode may be used in concert with the glass electrode,

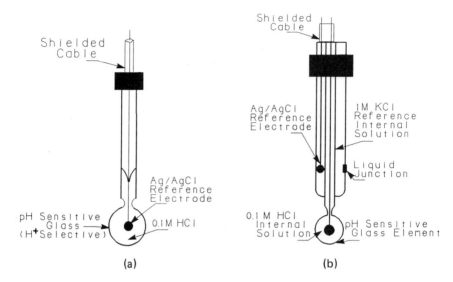

Figure 8.3   (a) Schematic of a glass pH electrode; (b) a combination pH electrode.

but the most common is probably the SCE, which is frequently used with the glass sensor in "combination electrodes" in which half-cells are combined in a single body (Figure 8.3b).

Apart from their utility for the measurement of the acidity of aqueous solutions, glass pH electrodes serve as the basis of measurement of a number of pollutant gases and find applications in monitoring their values in soils, water, and industrial emissions.

A schematic of a gas sensing electrode based on pH measurement is shown in Figure 8.4. It is, for all practical purposes, simply a combination pH electrode that is in contact with a very thin film of a solution whose pH may be modified by dissolution of an analyte gas. The solution film is held in contact with the combination electrode by a gas permeable membrane. Diffusion of the analyte gas through the membrane changes the acid–base equilibrium, and the extent to which the equilibrium shifts and the pH changes depends upon the partial pressure of analyte gas in the sample space. Gas sensing electrodes may be used as sensors in air or gas mixtures and in water.

Table 8.1 lists a number of pH based, pollutant gas sensing electrodes along with their sensor solutions, the equilibrium reactions involved, their detection limits, and some applications. It should be noted that use of an ammonia sensing electrode as a method for determination of ammonia in

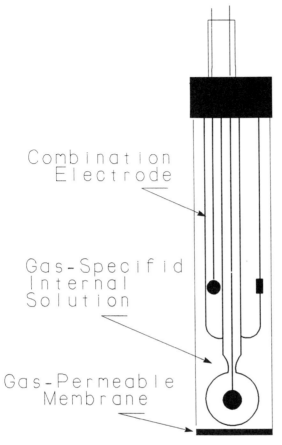

Combination
Electrode

Gas-Specifid
Internal
Solution

Gas-Permeable
Membrane

Figure 8.4 Schematic of a gas-sensing electrode — in this case making use of an internal combination pH electrode.

Table 8.1 Gas-Sensing Electrodes

Species	Internal electrolyte	Equilibrium reaction	Sensor electrode	Applications
NH_3	0.01M NH3	$NH_3 + H_2O = NH_4^+ + OH^-$	pH	Soil, organic N_2, water
CO_2	0.01M $NaHCO_3$	$CO_2 + H_2O = H^+ + HCO_3^-$	pH	Fermentation, blood gas
SO_2	0.01M $NaHSO_3$	$SO_2 + H_2O = H^+ + HSO_3^-$	pH	Stack gas, S in fuels
NO_2	0.02M $NaNO_2$	$2NO_2 + H_2O = NO_3^- + NO_2^- + H^+$	pH & NO_3^- ion selective	Stack gas, ambient air
H_2S	Citrate buffer	$H_2S + H_2O = HS^- + H^+$	S^{2-} selective	Pulps, fermentation
HCN	$KAg(CN)_2$	$Ag(CN)_2 = Ag^+ + 2CN^-$	Ag^+ selective	Wastes, plating baths

water is approved by the EPA [40 CFR Part 136.3] and is also an ASTM procedure [ASTM D 1426-79(D)].

Other potentiometric sensors used in environmental analysis include ion selective electrodes that are sensitive to H_2S and HCN. In both cases a silver sulfide membrane electrode is used. The electrode is constructed by sealing a pressed disk of polycrystalline silver sulfide (Ag_2S) onto a chamber containing a silver wire immersed in a silver nitrate solution. The electrode is specific to both silver and sulfide ions.

For the measurement of HCN, the Ag_2S electrode is placed in contact with a film of a solution containing potassium silver cyanide ($KAg(CN)_2$), which partially ionizes in solution according to

$$Ag(CN)_2 = Ag^+ + 2CN^-$$

So the diffusion of HCN through the gas permeable membrane into the solution film, forming H^+ and CN^- ions, drives the ionization reaction in the direction of the $Ag(CN)_2$ and, thus, reduces the concentration of Ag^+.

For measurement of H_2S, the electrode is placed in contact with a weakly acidic solution. Hydrogen sulfide diffusing into the weak acid dissolves and ionizes according to

$$H_2S = HS^- + H^+ = S^{2-} + 2H^+$$

the HS^- and S^{2-} ions changing the potential of the Ag_2S electrode.

Since the Ag_2S is specific to sulfide ions, it may also be used as a sensor for sulfides in water. Thus, it has been used to estimate sulfides, polysulfides, and sulfites in paper pulping liquors, in waste waters, and in sea water. It is able to detect sulfides at nanogram levels [Lakshminarayanaiah 1976, 173].

CONDUCTOMETRY

Electrolytic Conductivity

Like ordinary electrical conductivity, the conductivity of electrolyte solutions follows ohms law. That is, E = IR, where R is the resistance of the electrolytic cell. And, like the resistance of any conductor, R varies directly with the length of the electrolytic cell and inversely with its cross-sectional area. That is

$$R = \frac{\epsilon l}{a}$$

where ϵ is the specific resistance of the cell contents and l and a are the length and crosssectional area, respectively. The reciprocal of ϵ is known as the specific conductance, τ, so the conductance of an electrolytic solution is represented by

$$C = \frac{1}{R} = \frac{\tau a}{l}$$

where C has the dimensions of reciprocal ohms.

But the conductivity of an electrolyte solution also depends upon its concentration. So the equivalent conductivity, Ω, has been defined as the conductance of a solution containing one equivalent/cc of electrolyte that is confined to a cube with a volume of one cc. That is, Ω is defined as

$$\Omega = \tau/\text{equivalent/cc}$$

Measurements of the equivalent conductivity of electrolyte solutions show a strong dependence upon concentration, as is illustrated in Table 8.2. In addition, one finds that such measurements can distinguish two classes of electrolytes, strong and weak, with most salts and mineral acids being strong, or good conductors, and acids and bases like acetic acid and aqueous ammonia, respectively, being weak, or poor conductors, unless they are at very high dilution.

Extrapolating the values of Ω to zero concentration yields a value known as the equivalent conductivity at infinite dilution, Ω_0. As can be seen in Figure 8.5, that extrapolation may be made for strong electrolytes by plotting Ω as a function of $\sqrt{c}$, where c is concentration. Weak electrolytes, however, exhibit such rapid increases in conductivity at very high dilutions that measurement inaccuracies preclude determination of Ω_0 in that fashion.

Table 8.2 Equivalent Conductances of Some Electrolytes

Concentration	HCl	KCl	NaOH	AgNO$_3$	Acetic acid
0.0005	422.74	147.81	246	131.36	–
0.001	421.36	146.95	245	130.51	48.63
0.005	415.80	143.55	240	127.20	22.80
0.01	412.00	141.27	237	124.76	16.20
0.02	407.24	138.34	233	121.41	11.57
0.05	399.09	133.37	227	115.24	7.36
0.10	391.32	128.96	221	109.14	5.20

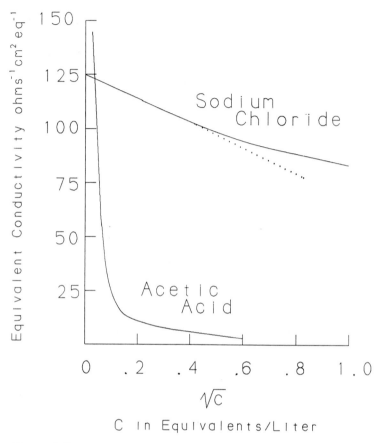

Figure 8.5 The concentration dependence of the equivalent conductances of strong (sodium chloride) and weak (acetic acid) electrolytes.

Equivalent conductivities at infinite dilution for a few electrolytes are shown in Table 8.3. Examination of that data reveals a very important characteristic of electrolyte conductivity. That is, when comparisons are made between salts having an ion in common, the differences in Ω_0

Table 8.3   Equivalent Conductances at Infinite Dilution

Electrolyte	Conductance	Electrolyte	Conductance	Difference
KCl	130.0	NaCl	108.9	21.1
KNO_3	126.3	$NaNO_3$	105.2	21.1
$\frac{1}{2}K_2SO_4$	133.0	$\frac{1}{2}Na_2SO_4$	111.9	21.1

Table 8.4 Ion Conductances at Infinite Dilution

Cation	Conductance	Anion	Conductance
H^+	349.82	OH^-	198.5
K^+	73.52	Br^-	78.4
NH_4^-	73.4	I^-	76.8
Ag^+	61.92	Cl^-	76.34
Na^+	50.11	NO_3^-	71.44
$\frac{1}{2}Mg^{2+}$	53.06	Acetate	40.9
$\frac{1}{2}Ba^{2+}$	63.64	$\frac{1}{2}SO_4^{2-}$	79.8

between them is a constant that is independent of the common ion. Apparently, each ion makes its own unique contribution to the total conductivity of the salt, regardless of the nature of its co-ion. This phenomenon is called law of the independent migration of ions and was discovered by Kolrausch in the latter half of the 19th century. Thus, because of independent migration, the equivalent conductivity at infinite dilution for any electrolyte is given by the sum of its individual ionic contributions

$$\Omega_0 = \Omega_0^+ + \Omega_0^-$$

where Ω_0^+ and Ω_0^- are called the limiting ionic conductivities.

Now recall from the previous section that the potential of a cell with a liquid junction is dependent upon the transference numbers of the migrating ions. Measurements of the potentials of such cells then, as a function of concentration, and extrapolation to infinite dilution, provide a way to determine transference numbers at infinite dilution. Since, by definition, transference numbers are the fractions of the total current carried by the electrolyte, i.e.,

$$t_{0^+} = \frac{\Omega_0^+}{\Omega_0} \quad \text{and} \quad t_{0^-} = \frac{\Omega_0^-}{\Omega_0}$$

one can combine their measurements with measurements of conductivity and isolate values for single ion limiting conductivities, some of which are shown in Table 8.4.

Conductivity as an Analytical Technique

Although lack of specificity and dependence upon concentraton limit the utility of direct measurements of electrolytic conductivity as an analytical tool, relative measures in conductivity can be put to advantage. For example, as will be seen, solution conductivity is used as a detection technique in ion exchange chromatography.

Differences in ionic conductivities, however, provide additional diversity to the method as an analytical tool. First, the equivalent conductivities of the ions at high concentrations bear the same relationship to one another that they do at their limiting values. That is, a good conducting ion at infinite dilution will be a relatively good conductor at higher concentration. The high values of the hydrogen and hydroxyl ions shown in Table 8.4 are, therefore, noteworthy, because they may be used to advantage in conductometric titrations, i.e, titrations in which the conductivity of a test solution is measured as a function of titrant additions. For example, if one titrates a strong base with a strong acid, the neutralization of the hydroxyl ion of the base by the hydrogen ion of the acid sharply reduces the conductivity of the base solution until all the base is gone. Beyond that point, continued titration brings excess hydrogen ions into the solution and sharply increases the conductivity of the test system. The point at which the change occurs is, of course, the acid-base equivalence point* and may be used to quantitate the amount of base that was in the test solution originally.

Conductometric titrations may be used to advantage in any situation in which an analyte ion with a low equivalent conductivity can be substituted for one with a high value, or vica versa. As will be seen later, this principle is also used to enhance eluent conductivity and, thus, the conductometric detection sensitivity to ions that have been eluted from ion exchange chromatographic columns.

Measurement of Electrolyte Conductivity

To avoid electrolyzing the test sample, electrolyte conductivity is always measured with an alternating voltage, usually at frequencies between 60 Hz and 1000 Hz.

For routine measurements, an electrode assembly is immersed directly in the test sample. The assembly is made of a pair of platinized platinum foil electrodes that are mounted within a glass cylinder (Figure 8.6a). Platinization is a process in which finely divided platinum is electrodeposited on the surfaces of the electrodes, which helps to inhibit electrolysis in the test solution. The glass cylinder serves to confine the flux of the electric current to a well defined geometry. This is necessary because

---

*The equivalence point is that point at which the number of equivalents of added reagent is exactly equal to the number of equivalents of analyte in the test solution.

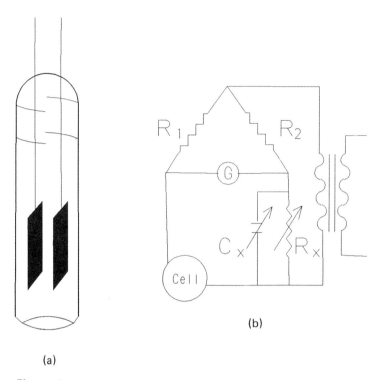

(a)

(b)

Figure 8.6   (a) A cell for the measurement of solution conductance; (b) a typical measuring circuit for solution conductivity.

of the current paths that are possible in electrolyte solutions. That is, if one maps the current lines, it is seen that they behave very much like magnetic flux lines do about the poles of a magnet. They are most dense in paths normal to the planes of the electrodes and exhibit decreasingly dense paths from one electrode to the other as a function of the distance away from the electrodes. But they distribute, thereby, throughout the entire volume of a test solution. Thus, electrodes not confined to a fixed volume would exhibit conductivity values dependent upon the geometry and the volume of the test vessel.

Commercial assemblies come in several conductivity ranges that are determined by the surface area of and the spacings between the platinum foil electrodes.

Solution resistance is measured most frequently by means of a Wheatstone bridge as is shown in Figure 8.6b. Voltage is applied across a pair of dividers comprised of two equal resistances, R_1 and R_2, the test solution, and a variable resistor R_x. An AC galvanometer is connected between the midpoints of the divider networks. When R_x is adjusted to zero current through the galvanometer, its resistance is equal to the test sample's.

Because the parallel plate electrodes are equivalent to a resistor and a capacitor in parallel, and exhibit both resistive and capacitive reactance when an alternating voltage is impressed on them, a variable capacitor is placed in parallel to the measuring resistor in the bridge to balance the capacitance.

A given electrode assembly is calibrated by means of a solution of known equivalent conductivity. A conductivity "cell constant" is thus determined which is given by

$$k = \frac{\Omega_{known}}{C_{known}}$$

and the unknown's equivalent conductivity is obtained from

$$\Omega_{unknown} = kC_{unknown}$$

where C refers to the measured conductivity values.

Applications of Conductometry

EPA guide lines prescribe field and laboratory conductivity measurements as a check for sample degradation when sampling for acid rain studies [Operations and Maintainance Manual 1986, Part 1, Section 7.0, 5]. And conductivity, as mentioned earlier, is used as a detector for ion chromatographic analysis, with applications specific to our subject elements including determination of ammonium, nitrate,and sulfate ions in acid rain.

KINETIC METHODS

Principles

In contrast to potentiometry, in which measurements are made under conditions of zero current flow, so as to avoid changing the chemistry of the cell components, other electrochemical methods make use of the current flow associated with redox reactions as a quantitation vehicle. We are defining this class of methods as kinetic.

Recall the distinction between galvanic and electrolytic cells. The former is a source of current and voltage, generated by spontaneous redox reactions underway within it. The latter cell is one in which redox reactions are induced by the application of an external voltage. Potentiometric cells are galvanic. However, if an external voltage with opposite polarity and magnitude somewhat greater than its normal galvanic potential is applied to it, a potentiometric cell becomes electrolytic i.e., its galvanic electrode reactions are reversed in direction. The reversal of the redox processes in the cell is called electrolysis, and electrolysis is the basis of a number of kinetic electrochemical analytical methods.

Our discussion of kinetic methods will be confined to two general approaches in which electrolysis is used for quantitative analysis in environmental applications; these are coulometry and polarography (especially an application of polarography called amperometry). In the first, the *amount* of electricity required to quantitatively oxidize or reduce an analyte is related to its concentration. In the second, the *current flow* through an electrolytic cell in which a redox reaction is underway is related to analyte concentration.*

To explain kinetic methods, it is necessary to first describe the behavior of electrolytic cells when a voltage is applied to them. A typical cell is shown in Figure 8.7. It is composed of a rotating platinum electrode immersed in a test chamber that is connected to a reference electrode through a salt bridge. The physical size and the concentration of the internal solution of the reference electrode are such that passage of current through it will not change its half-cell potential. The voltage applied to the cell is made negative with respect to the reference electrode, thus causing a reduction process to occur in the test chamber.

When increasingly negative voltages are applied, several transitions in the behavior of the current flowing through the cell occur; these are shown in Figure 8.8. At low voltage levels, the electroactive components in the cell migrate toward the electrodes of charge opposite to their own. Electrical neutrality then causes the formation of layers of ions of opposing charge extending from the immediate neighborhood of the electrodes out into the bulk of the solution. The structure of these "double layers" becomes progressivley more diffuse as a function of the distance from the electrodes. But since convective and diffusive forces oppose the formation of any ionic ordering, a small *residual current* flows through the cell to counter double layer dispersion.

As the voltage is increased beyond the normal galvanic potential of the cell, the electroactive component in the test chamber begins to undergo

---

*The distinction between amperometry and conductometry is that in the latter, current is measured in a system that is not undergoing oxidation/reduction while in the former it is.

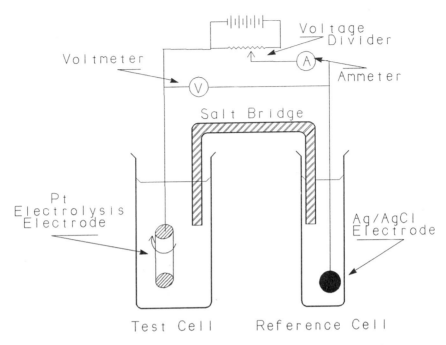

Figure 8.7   Schematic of an experimental set-up for developing a polarogram.

reduction, and the current increases. The potential at which this occurs is called the *decomposition potential*. Increasing voltage is then accompanied by increasing current as long as the electroactive species can migrate to the electrode in sufficient quantity to pace the voltage. However, a point is ultimately reached when the discharge of the electroactive species is occurring faster than they can migrate to the electrode surface. That is, the current becomes diffusion limited, and levels off to a value called the *limiting current*, and the concentration of the electroactive species at the electrode surface approaches zero.

Thus, at the value of the limiting current, a concentration gradient is developed between the electrode surface and the bulk of the test solution that may be expressed as

$$\frac{dC}{dx} = \frac{C_b - C_0}{\delta}$$

where C_b is the concentration in the bulk solution, C_0 the concentration

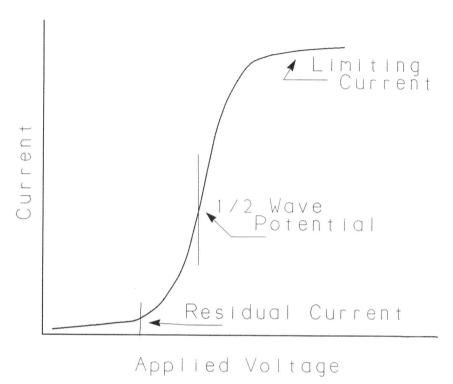

Figure 8.8   A polarogram for a single reducible component.

at the electrode surface, and δ the thickness of the gradient.*

From Fick's law, however, we know that the rate at which diffusing substances pass through a concentration gradient is given by

$$\frac{dN}{dt} = AD \frac{dC}{dx}$$

where D, the proportionality constant, is called the diffusion coefficient, A is the cross-sectional area of the gradient, and dN/dt is the number of moles of diffusing species passing across the gradient per unit time. So, combining the two relationships and setting C_0 equal to zero, one gets

$$\frac{dN}{dt} = \frac{ADC_b}{\delta}$$

---

*A rotating electrode is used to both minimize and control the value of δ.

And, since the current is related to the passage of the electroactive components across the gradient, the current due to their diffusion is given by

$$i_d = k \frac{dN}{dt} = \frac{kADC_b}{\delta} \qquad (8.13)$$

and i_d is seen to be directly proportional to the bulk concentration of the
reducible component of the test solution.

Polarography and Amperometry

Current–voltage curves of the kind shown in Figure 8.8, and the relationship of the limiting current to concentration, serve as the foundations of
kinetic electroanalytical methods. The value of the voltage at the mid-
point between the residual current and the limiting current, which is called
the *half-wave potential*, for example, may be shown to be nearly equal to
the standard potential for the redox reaction underway in the cell [Picker-
ing 1971, 474] Thus, the half-wave potential may be used to suggest the
identity of the electroactive entity in the test solution, and the limiting
current provides its concentration.

In mixtures of redox couples whose standard potentials are sufficiently
separated, current-voltage curves appear as a series of plateaus (Figure
8.9). Each of the plateaus corresponds to the limiting current of one of
the redox couples being reduced at the electrode, and the mid-point
between the plateaus is the value of that couple's half-wave potential.
Thus, a mixture of electrolytes can be quantitatively analyzed by developing such curves, which are called *polarograms*; the analytical method
involved is called *polarography*.

Amperometry is a special case of polarography in which the applied
voltage is set to a level that is high enough to decompose the analyte as
well as to maintain a current flow equal to the limiting current over the
expected analyte concentration range. The appropriate voltage level is
determined experimentally for any given redox system. The measurement
of the cell current can then be related to the concentration of the electroactive component of the test solution.

Similar to conductometry, amperometry may be used as an equival-
ence-point detector in titrations. That is, the limiting current exhibited
by a solution of electroactive material will decrease if its concentration is
decreased by the addition of a reacting titrant. The equivalence point is
signalled by the reduction of the cell current to its residual level.

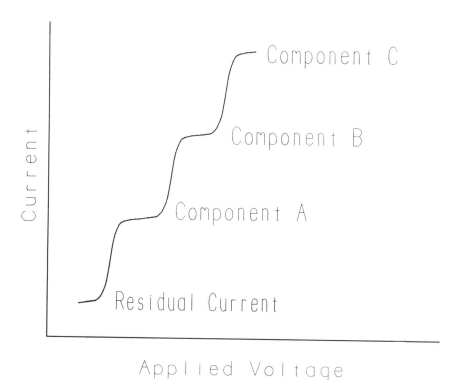

Figure 8.9   A polarogram for a mixture of reducible ions with standard redox potentials well separated.

Coulometry

In coulometry, in contrast to amperometry, current flow is integrated over the time interval required to either reduce the concentration of an analyte to zero or to generate a reactant that quantitatively reacts with the analyte being determined. Thus, for a generalized redox reaction of the kind

$$A^{a+} + ne^- = A^{(a-n)+}$$

where a is the ionic charge and n is the number of electrons involved in the reduction process, the amount of electricity passed to quantitatively reduce A^{a+} is given by

$$Q = \int idt = \frac{nwF}{M}$$

and the amount of material reduced is

$$w = \frac{M}{nF} \int idt$$

where M is the molecular weight of the substance, w is its weight, and F is the Faraday.

The end point of a coulometric determination may be signalled by the reduction of the current flow in a coulometric cell to the level of the residual current for the system. Alternatively, a coulometric titration may be conducted in which a titrant is electrically generated rather than being added volumetrically, the equivalence point being determined by an amperometric detector (see below).

Applications of Kinetic Methods

An important and EPA approved application of amperometry is its use for the determination of biochemical oxygen demand (BOD) by means of an amperometric oxygen electrode [40 CFR, Part 136.3]. Similar to potentiometric gas sensing electrodes, the amperometric oxygen electrode [Clark 1956] makes use of a gas permeable membrane that allows oxygen to diffuse to an oxygen sensing element (Figure 8.10). The oxygen element is constructed from a platinum wire immersed in a potassium chloride solution that is negatively polarized with respect to a Ag/AgCl electrode residing in the same solution. The voltage between the electrodes is fixed between -0.6 and -0.8 volts, a value sufficient to maintain limiting current conditions for the reduction of oxygen to OH^- ions. The diffusion rate of the oxygen across the membrane and, thus, the level of the limiting current are proportional to the oxygen concentration in the test sample.

An EPA approved liquid chromatographic procedure for the determination of benzidine and dichlorobenzidine in water makes use of a similar amperometric electrode for detection and quantitation of the chromatographic eluent [40 CFR, Part 136, Appendix A].

Carbon and sulfur, as carbon dioxide and sulfur dioxide, respectively, are quantitated by coulometric titration in a number of commercial combustion analyzers (see below in section on chemical methods). In the case of carbon, the carbon dioxide generated by combustion of carbon-containing samples is passed into a solution of water, dimethyl sulfoxide, and ethanolamine, where it is absorbed to form 2-hydroxyethylcarbamic acid. The acid is detected photometrically by means of an acid-base indicator also incorporated into the test solution. To quantitate the CO_2, the acid is neutralized by base generated from the electrolysis of the water

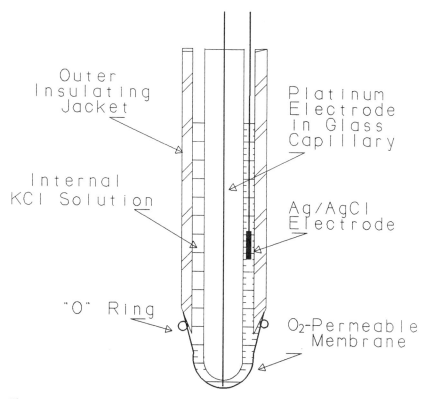

Figure 8.10   Schematic of and amperometric oxygen electrode (after Clark).

in the test solution, and the amount of electricity consumed to neutralize the acid is related to the carbon dioxide absorbed.

Sulfur dioxide may be quantitated similarly by passing it into a solution of free iodine and an iodide salt. The sulfur dioxide reacts with the iodine and water according to

$$SO_2 + I_2 + H_2O = H_2SO_4 + 2I^- + 2H^+$$

and the reduced level of iodine or the increased level of iodide ion is sensed by an indicator electrode. The signal from the indicator controls the electrolysis of the test solution, which regenerates free iodine from the iodide being formed in the reaction. Quantitation of the SO_2 is through measurement of the quantity of electricity required to generate enough iodine to react with all the SO_2 passing into the test solution, a point that is determined by the appearance of excess free iodine.

Both potentiometric and amperometric indicator electrodes may be used in these "iodometric" coulometric titrations. Potentiometric indicators are made from silver or silver/silver iodide elements, which are reversible to iodide ion and generate a potential related to its concentration in the test solution. They are generally paired with calomel references.

Iodometric amperometry, however, operates on a somewhat different principle than that discussed previously. In contrast to our earlier discussion, a pair of platinum indicator electrodes is used. Between these electrodes is fixed a potential *below* the decomposition potential of the iodine/iodide redox couple. In such a system, when iodine is absent, only residual current may flow: the system behaves like a high resistance. But if iodine is present, the system resistance decreases and higher current levels can flow, despite the fact that the applied potential is lower than the decomposition potential.

The decrease in resistance is a consequence of the symetry of the half-cell reactions that may take place in the presence of free iodine

$$2I^- = I_2 + 2e^- \quad \text{and} \quad I_2 + 2e^- = 2I^-$$

Since the electrode reactions are symetric, there is no net free energy change associated with them when current flows. Recalling our earlier discussion, we know that the galvanic potential of such a system is zero. Thus, when iodine is present, the applied voltage required to cause current to flow is only that necessary to overcome the ordinary resistance of the test solution, i.e., the resistance associated with the equivalent conductances of its components [Lingane 1953, 388].

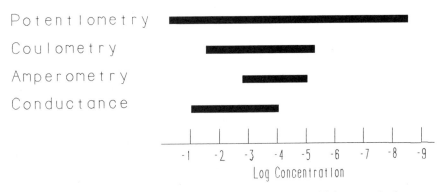

Figure 8.11 Approximate concentration ranges over which several electrochemical methods are applicable.

SPECIFICITY AND SENSITIVITY OF ELECTROCHEMICAL METHODS

As was mentioned before, electrochemical methods are not particularly specific, current and voltage characteristics being dependent, in general, on the overall composition and concentration of the components of an analysate. Thus, their greatest strength lies within the scope of quantitating the components of systems of known composition. But electrochemical methods do have high sensitivity, being able with some methods to quantitate analyte levels as low as 10^{-8} or 10^{-9} molar. Figure 8.11 illustrates the approximate ranges over which a number of electrochemical methods are applicable.

9

Chromatography

PRINCIPLES

Similar to spectroscopy, the term chromatography is generic in the sense that it is applied to a number of distinctly different analytical techniques that share a common physical principle. In its several forms, chromatographic analysis is, perhaps, one of the most important and powerful of the analytical techniques applied to environmental problems. It is used extensively for the determination of specific gaseous pollutants in the air and is an EPA approved methodology for the detection and determination of a large number of organic pollutants in water.

As noted earlier, chromatography is fundamentally a separation science. Its origin as an analytical method lies in work of the Russian botanist Mikhail Tswett, who, in the early part of the century, separated an extract of plant pigments into its components by passing them through a glass column packed with powdered calcium carbonate. As the mixture passed through the column, the pigments separated into a set of colored bands, from which the method's name is derived, i.e., from the Greek *chromatus* and *graphein*, meaning "color" and "to write," respectively.

Modern chromatography, of course, encompasses a much broader range of techniques, only a few of which are based on color separation, but all of which are based on the same operational principles that separated the plant pigments in Tswett's experiments. That is, chromatography now refers to any separation technique in which separation is effected by distribution of a sample between two immiscible phases, one of which is mobile and the other stationary. The mobile phase may be a gas, a liquid, or a super-critical fluid,* and the stationary phase may be a solid or a liquid film that has been immobilized on a solid by adsorption or chemical binding. When the mobile phase is a gas, the method is referred to, generically, as GC or gas chromatography; if it is a liquid, the method is called liquid chromatography. More specific designations define the stationary as well as the mobile phases. Thus, LSC, for example, refers to liquid-solid chromatography, meaning a liquid mobile phase in combination with a solid stationary phase, and so forth.

In practice, the mobile phase is passed over or through the stationary phase. Separation is implemented by injecting a mixture of analytes into the mobile phase,† upstream of the stationary phase. The separation then occurs as a consequence of the extent to which the analytes equilibrate between the mobile and stationary phases. That is, when the analyte molecules are introduced into the mobile phase and are carried through the system, some of them "stick," so to speak, to the stationary phase, causing them to distribute between the two phases in ratios that depend upon the relative strengths of their interactions with each. As fresh, analyte-free, mobile phase follows and travels by "stuck" molecules, it establishes new equilibria between the two phases and, thus, "unsticks" molecules from upstream positions and carries them further downstream where the process repeats itself. Molecules are, therefore, continually moving between the mobile and stationary phases, with those that are strongly attracted to the stationary phase being retarded in their transit through the system to a greater extent than those that are weakly attracted to it. As a result, the components of the mixture separate as they are carried though the system, and the separation between them becomes greater as a function of their transit distance.

In standard gas and liquid chromatography, the stationary phase is packed into a column through which the mobile phase is passed at a constant flow rate (column chromatography). One special case of liquid chromatography, however, makes use of a stationary phase coated onto a length of a chromatographically inert flat surface like a piece of glass. This special case is called thin layer chromatography (TLC).

In column chromatography, the mobile phase is used to effect complete elution of the analyte from the column, the presence of which in the eluent may be detected by any of several methods that will be discussed

---

*A liquid and its vapor are always in equilibrium at normal temperatures and pressures. As temperature and pressure is increased, however, a point is reached, called the critical point, where the liquid and vapor phases become indistinguishable. The state of matter above the critical point is called a super-critical fluid.
†One selects a mobile phase in which the analytes are soluble.

below. The time required to effect elution of the analyte is called its *retention time*, and the volume of mobile phase associated with the retention time is called the analyte's *retention volume*. Since common practice in column chromatography is to maintain mobile phase flow at a constant rate, the retention volume is simply the product of the flow rate and the retention time. Figure 9.1 is an illustration of a typical column chromatogram and shows the relationship of the retention time and volume to the position of an eluting analyte band. The area of a band or its peak height may be related to an analyte's concentration.

In thin layer chromatography, the analysate is deposited at one end of the surface coating, and the mobile phase is allowed to move by capillary action, carrying analyte molecules across the surface. Thin layer chromatography is, therefore, closely related to Tswett's original experiments in that the separated analytes are not eluted from the thin layer plate but simple separated upon it. They are then usually detected by treating the plate with some kind of color development reagent.

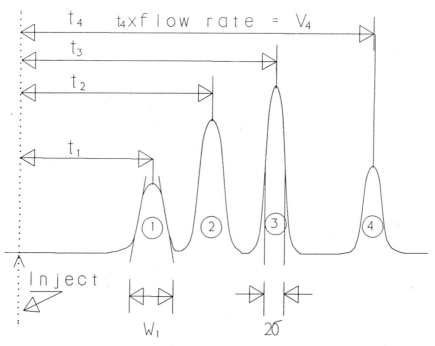

Figure 9.1   A column chromatogram showing how retention time, retention volume, the peak width, and the peak half width are defined. Note that if the peaks were truly normal distributions, σ would represent one standard deviation.

Depending upon the nature of the stationary phase, its interaction with the analyte may take two forms; it may physically adsorb it to its surface or it may absorb or dissolve it. With solid stationary phases, adsorption is usually the interaction mechanism, and chromatographers call that process *adsorption chromatography*. With liquid stationary phases, solution is usually the interaction mechanism, and the process is generally called *partition chromatography* because of the "partitioning" of the analyte between two solution phases.

Both solid and liquid stationary phases are also used in a special application called *exclusion chromatography*. In exclusion chromatography, the stationary phases are made from porous solids or liquid gels, which have been chemically synthesized in such a way as to limit their pore sizes to well defined ranges. Separation in exclusion chromatography is based on differences in the sizes of the analyte molecules present in the analysate. That is, only those molecules that physically fit into the pores of the stationary phase are retarded as the sample is transported through the system.

Some Definitions and Quantitative Relationships

The distribution of analyte between the mobile and stationary phases is given by a *partition coefficient* defined as

$$K = \frac{C_s}{C_m}$$

where C_s and C_m are the equilibrium concentrations of the analyte in the stationary and mobile phases, respectively. The magnitude of the partition coefficient determines the extent to which analyte is retarded by the system, while differences in partition coefficients between analytes determine the extent to which they will separate.

The relationship between the partition coefficient and the degrees of retardation and separation may be inferred from the following arguments. In their transit through a chromatographic system, analyte molecules are continually exchanging between the mobile and stationary phases. However, they may progress through the system only when they are part of the mobile phase.* Their retention times, therefore, must be the sum of the times they spend in each of the phases. That is

$$t_r = t_m + t_s$$

---

*Clearly, all analyte molecules must spend equal amounts of time in the mobile phase as they progress through the system.

where t_r is the retention time, t_m is the time spent in the mobile phase, and t_s is the time spent in the stationary phase.

On average, then, the fraction of time analyte molecules spend in the mobile phase is given by

$$R = \frac{t_m}{t_r} = \frac{t_m}{t_m + t_s} \tag{9.1}$$

where R is defined as a *retardation factor*.

Now, because of the partition coefficient, there is always a fixed fraction of analyte molecules in the mobile phase as the analyte zone traverses the system. The likelihood of any single analyte molecule being in the mobile phase is equal to the fraction of analyte residing there, and the fraction of the retention time that each molecule spends in the mobile phase is equal to the likelihood of its being there. Therefore, since all analyte molecules must spend equal times in the mobile phase as the zone traverses the system, one can write that the fraction of the retention time that analyte as a whole spends in the mobile phase is equal to the fraction of it that resides there. That is

$$\frac{t_m}{t_m + t_s} = \frac{\text{mass in mobile phase}}{\text{total mass}}$$

Using similar arguments for the stationary phase, one can also write

$$\frac{t_s}{t_m + t_s} = \frac{\text{mass in stationary phase}}{\text{total mass}}$$

Dividing these two expressions yields

$$\frac{t_m}{t_s} = \frac{\text{mass in mobile phase}}{\text{mass in stationary phase}}$$

and defining V_s and V_m as the volumes of the mobile and stationary phases, respectively, one gets

$$C_m V_m = \text{mass in the mobile phase}$$

and

$$C_s V_s = \text{mass in the stationary phase}$$

Thus,

$$\frac{t_m}{t_s} = \frac{C_m V_m}{C_s V_s} = \frac{1}{K}\frac{V_m}{V_s}$$

and the retardation factor becomes

$$R = \frac{1}{1 + K(V_s/V_m)} \qquad (9.2)$$

Since analyte molecules are swept along by mobile phase, their velocity, while in the mobile phase, must be equal to that of the mobile phase molecules. Since both must traverse the same transit distance, it is clear that the time analyte molecules spend in the mobile phase must be equal to the time it takes mobile phase molecules to traverse the system. Then, from eq. 9.1

$$R = \frac{t_m}{t_r} = \frac{t_0}{t_r} \qquad (9.3)$$

where t_0 is the transit time of mobile phase molecules. Combining eqs. 9.2 and 9.3 and rearranging then yields

$$t_r = t_0\left(1 + K\frac{V_s}{V_m}\right) \qquad (9.4)$$

and it seen that the retention time is directly proportional to the partition coefficient.

From eq. 9.4, we may also see that the separation between two analytes, a and b, is given by

$$t_{ra} - t_{rb} = t_0 \frac{V_s}{V_m}(K_a - K_b)$$

Thus, the larger the differences between their partition coefficients, the better will be the separation.

Band Broadening

As analytes are carried down a chromatographic system by the mobile phase and separate, each of the separated components forms a band with a shape approaching that of a normal distribution, with the ordinate representing concentration and the abscissa representing time or distance flowed. As separation continues with increased transit distance or time, however, the individual bands broaden and decrease in height. Thus, in addition to the partition coefficients, there are other factors that enter into the efficiency of a chromatographic separation. Three major contributors affect band broadening, two of which are dependent upon mobile phase velocity and the third dependent upon the physical configuration of the stationary phase. The velocity dependent contributions are due to mass transfer in the column and molecular diffusion within the analyte band. The third contribution results from the way the mobile phase flows through the stationary phase and is called eddy diffusion.

Mass transfer broadening is a consequence of the continuing distribution and redistribution of analyte between the stationary and mobile phases. That is, as molecules distribute between the two phases in their passage through the system, those remaining in the mobile phase move some distance downstream from those that are left behind in the stationary phase. This is illustrated in Figure 9.2, which shows the displacement of the analyte concentration profile in the mobile phase from that in the stationary phase. Because of the displacement, analyte from the leading edge of the mobile phase redistributes into the stationary phase beneath

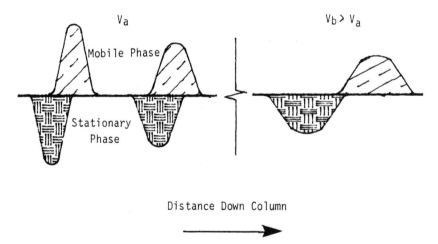

Distance Down Column

Figure 9.2   The displacement of analyte in the mobile phase from that in the stationary phase as the front moves down a chromatographic column. An increase in mobile phase velocity will emphasize the effect, as is shown in the right side of the figure, and increase mass transfer band broadening.

it, and analyte in the stationary phase redistributes into the trailing edge of the mobile phase above it, the net effect being to broaden the overall zone in which the analyte is concentrated. As can be seen in Figure 9.2, band broadening due to mass transfer increases with increased mobile phase velocity because of the greater displacement of the residual analyte in the mobile phase from that in the stationary phase in any increment of time.

Band broadening from molecular diffusion is illustrated in Figure 9.3 and arises because the laws of diffusion dictate that molecules move from regions of high to low concentration. Thus, they migrate from the high concentration region at the center of the analyte band toward its boundaries. Molecular diffusion, of course, is always underway within an analyte band as it is carried through the system by the mobile phase. Thus, its effect is opposite to that of mass transfer in terms of mobile phase velocity; that is, high mobile phase velocities favor minimum broadening due to molecular diffusion.

Eddy diffusion is a consequence of the tortuous flow paths that the mobile phase follows in moving through a typical chromatographic system containing a stationary phase made from packed granular particles. Clearly, the mobile phase and the analyte molecules it carries take the paths of least resistance in their transit through the system. As a result, the downstream flow is in the form of randomly directed eddies, which

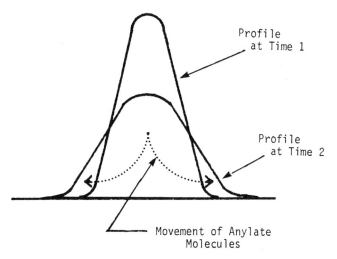

Figure 9.3   Band broadening due to diffusion of analyte molecules from the high concentration at the center of the band to the lower levels at its periphery.

have the effect of spreading the analyte molecules over broader lengths of the stationary phase. Eddy diffusion is also flow rate dependent but is considered to be a minor contributor to band broadening at normally used mobile phase flow rates [Miller 1988, 31].

Because of these various effects and their dependence upon mobile phase flow rate, however, designs of chromatographic systems are always a compromise of transit distance or column length, column geometry, the state of subdivision of the stationary phase, and the mobile phase velocity, so as to optimize separation efficiency while minimizing band broadening and the time required for the separation to be analytically effective.

The Quality of a Chromatographic System

There are three figures of merit that are used to describe the behavior of chromatographic systems; these include column selectivity, column efficiency, and column resolution.

Selectivity is defined as the ratio of the partition coefficients of two analytes on a particular chromatographic system. The ratio is called the *separation factor* and is given by

$$\alpha = \frac{K_b}{K_a}$$

where the K's are the partition coefficients. Selectivity is clearly a function of the chemical and physical interactions between the analytes and the mobile and stationary phases.

Column efficiency is a measure of the column's ability to produce distinct analyte bands that are well separated from one another and not so broad that they are hard to distinguish from a base line, i.e., from a detector signal generated by analyte-free mobile phase passing through it. Column efficiency is a function of selectivity, the column configuration, and the mobile phase velocity. It is usually expressed in terms of a *plate number* or the number of theoretical plates* the column contains and is given by

$$n = \left(\frac{t_r}{\sigma}\right)^2$$

---

*The term "plate number" in chromatography is a carry-over from early theoretical treatments in which separation on chromatographic columns was considered to be analogous to distillation. In distillation theory, separation efficiency is given by the number of theoretical plates of the distillation column.

where σ is half the width of the analyte band at a position representing its half height (Figure 9.1).* We see therefore, that the plate number represents a measure of band broadening as a function of retention time, implying that a column with a high plate number produces very narrow, rapidly eluting analyte bands. That is, the higher the value of the plate number, the more efficient the resolving power of the column.

Retention time, however, is also a function of the length of the chromatographic column, so n also depends upon length. Thus, an alternative definition of efficiency is called the *height equivalent of a theoretical plate*, or the plate height, and is given by

$$H = \frac{L}{n}$$

where L is the length of the column; the lower the value of H, the more efficient is the column.

Of course, the most important quality of a chromatographic system is its ability to separate or resolve the bands representing the components of an analysate. An empirical measure of resolution, therefore, has been defined in terms of the separation between the two closest bands of a chromatogram and is given by

$$\text{Resolution} = 2\frac{t_{ra} - t_{rb}}{w_a + w_b} \tag{9.5}$$

where t_{ra} and t_{rb} are their retention times and w_a and w_b are the band widths at their bases. Clearly, large values of Resolution are desirable, and bands that are completely separated exhibit Resolution values of 1.5 or larger [Miller 1988, 18].

For bands with approximately the same widths, eq. 9.5 reduces to

$$\text{Resolution} = \frac{\delta t_r}{w}$$

But from our discussion above, we know that δt_r may be increased by increasing the column length, i.e., transit distance. In fact, it can be shown [Hamilton & Sewell 1982, 29] that

$$\delta t_r \sim \text{transit distance}$$

---

*If the chromatographic bands were true normal distributions, σ would correspond to the standard deviation.

and that

$$w \sim \sqrt{\text{transit distance}}$$

Therefore, resolution may always be increased by extending the length of the chromatographic column, but at the expense of both analysis time and line broadening.

Equation 9.5 is useful to characterize any particular chromatographic column; it serves as a quality indexing device. It provides no insights into how column behavior might be optimized, however. But in theoretical modeling studies [Purnell 1960], it has been shown that resolution between two adjacent bands can be quantitatively related to the retention time, plate number, and selectivity factors that have been discussed above. Thus, Purnell has shown that

$$\text{Resolution} = \frac{1}{4}\sqrt{n_2}\,\frac{(\alpha - 1)k_2}{\alpha(1 + k_2)}$$

where the subscript denotes the second of two analytes in the chromatogram, n is the plate number, α is the selectivity, and k is defined as

$$k = \frac{t_r - t_m}{t_r}$$

We see, therefore, confirmation of the influence of the plate number on resolution along with the contributions from factors describing the interactions of the analytes with the mobile and stationary phases.

Temperature Control

Temperature control in gas and liquid chromatography is essential because the distribution of analyte between the stationary and mobile phases represents an expenditure of thermodynamic work, i.e., a change in free energy. The free energy change involved is similar to that given by eq. 8.2. That is,

$$\delta G = \delta G_s - \delta G_m = RT(\log[C_s] - \log[C_m])$$

and

$$\log K = \frac{\delta G}{RT} \tag{9.6}$$

Therefore, the higher the temperature, the lower the value of the partition coefficient, and the lower is the retention time. Because of temperature sensitivity, and whether one is dealing with LC, GC, or their various submethods, it is good practice to thermostat chromatographic columns. However, commercial liquid chromatographic instruments are frequently run at ambient temperatures, the value of δG for liquid systems being lower than that for gas systems.

Chromatographic Analysis

From the foregoing discussion, it is clear that the behavior of chromatographic columns (and TLC plates) will vary from one to the next. Systems will also exhibit some variability because detectors do not necessarily respond in exactly the same way to all analytes. Thus, in practical applications of chromatography, reference chromatograms for known materials are first developed. Identity and quantitative information about unknowns are then obtained by referring test chromatograms to the references. Unequivocal identification of unknowns can not generally be made except by reference to known materials because retention times, though unique to each analyte passing through a given system, are not specific. That is, in a complex mixture, it is possible for two components to exhibit similar retention times.

GAS CHROMATOGRAPHY

Equipment in General

Figure 9.4 is a schematic of a gas chromatograph, showing its essential elements. The instrument consists of a source of gas to serve as the mobile phase, a flow controller to maintain a constant mobile phase flow rate, the column packed with stationary phase, a sample injection device, and a detector.

Gas chromatographic columns may be run isothermally or programmed to increase in temperature as a function of time. Temperature programming allows for more rapid separation of analytes with widely varying partition coefficients. That is, as temperature is increased, the partition coefficients for analytes that interact very strongly with the stationary phase are reduced (eq. 9.6). Thus, raising the temperature of the column reduces their retention times and allows them to be eluted in the same relative time frame as their co-analytes with lower partition coefficients.

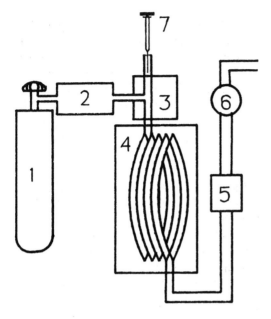

Figure 9.4 The components of a gas chromatograph: (1) the carrier gas source; (2) gas flow regulator and controller; (3) thermostated sample injection port; (4) thermostated column; (5) detector; (6) a gas flow meter; (7) a syringe injector.

The Mobile Phase

Since, in gas chromatography, the only function of the mobile phase is to sweep analyte vapor molecules through the column,* it is called a carrier gas, and the primary criteria for its selection is that it be chemically and physically inert with respect to both the analyte and the stationary phase. The most frequently used gases are highly purified nitrogen, hydrogen, and helium. As will be seen, selection of one or the other of these is frequently dictated by the kind of detector that is used.

Sample Injection Devices

Several different kinds of sample injection devices may be used. The most common and least expensive is a well calibrated, gas-tight syringe that may be used to inject sample through a self-sealing elastic diaphragm

---

*As will be seen, in liquid chromatography, mobile phases are selected for their solvent properties relative to the analytes.

called a septum. Sliding gas valves that capture a fixed volume of flowing sample in a channel and then "slide" the channel into the carrier gas plumbing circuit are also used. And, finally, so-called *purge and trap* devices may serve as direct input systems. Most sample injection systems are heated to assure that analyte samples are in the vapor state upon their entry into the column.

All sliding gas valves are based on principles similar to that shown in Figure 9.5. The center, circular section of the valve rotates within the hexagonally shaped external section. Internal channels make connections as shown. In standby position, they link a syringe to a sample loop made from a small length of tubing connected to the valve. Sample size may thus be adjusted by connecting different lengths of tubing into the sample loop. Simultaneously, another internal channel connects the source of the mobile phase to the column. To inject the sample into the column, the center section of the valve is rotated 60°, switching, as can be seen, the sample loop into the mobile phase circuit while isolating the syringe.

Although syringe injection through a septum is the most common

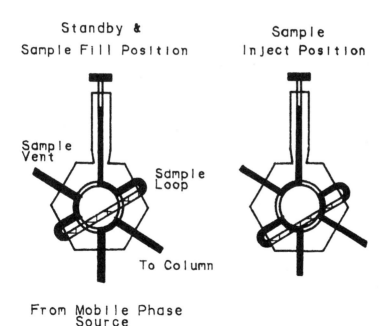

Figure 9.5   A chromatographic sample injection valve. In the standby position, the sample loop is filled. In the inject position, the loop is switched into the carrier gas circuit.

sample input mode used for gas chromatography, valves like the one illustrated in Figure 9.5 are better for careful quantitative analysis because their mode of action captures a very precise volume element and provides for very reproducible sample injections.

Purge and trap systems are essentially temperature controlled adsorption columns. They are used primarily for analysis of extremely low level analytes. A large volume of analysate is collected and purged or swept by an inert gas into the system where it passes through a column (a trap) that contains a high surface adsorbent like activated charcoal. The analyte is adsorbed and, thus, concentrated on the adsorbent. By raising the temperature of the trap, the analyte may then be driven off it and swept into a chromatographic column.

Purge and trap systems may be valved with slide valves similar to the kind shown in Figure 9.5, the sample loop being connected between the purge and trap elements, i.e., between the analysate source and the adsorption column. After purging the sample from its source and into the adsorption trap, the temperature is raised, and the valve is rotated so as to link the trap into the carrier gas circuit of the chromatograph.

Columns

There are two kinds of gas chromatographic columns, packed and open tube. Packed columns may be made of glass, metal (stainless steel, copper, or aluminum), or teflon tubing, while open tube columns are most frequently made of glass or fused silica capillary tubing. Packed columns are of the order of a few millimeters in diameter and 1–2 meters in length. Open tube columns are of the order of 0.25 mm in diameter and may be as long as 60 m in length.

Packed columns contain stationary phases made from finely divided solids. In the case of GSC, the packing itself serves as the stationary phase. In the case of GLC, the packing serves as an inert solid support for the liquid stationary phase. Typical solid column packings include such materials as molecular sieves,* zeolites, diatomaceous earth, silica gel, alumina, glass beads, and charcoal.

Liquid stationary phase materials are selected for specific separation applications, a general selection rule being that they have solvent properties for the analytes being separated. Thus, polar analytes are best separated on polar stationary phases, etc.

Possibly the most commonly used liquid stationary phases are those

---

*Molecular sieves are synthetic porous solids that contain pores with well defined size ranges. They are used primarily for gas-phase exclusion chromatography.

based on a family of liquid silicone polymers, the backbone chemistry of which is

$$\left[\begin{array}{c} R \\ | \\ -Si-O- \\ | \\ R \end{array} \right]_n$$

where R designates a substituent group. By proper selection of the substituent groups, liquid stationary phases of varying polarity may be synthesized and used for the separation of similarly varying polar vapors.

Other typical stationary phase materials include long chain hydrocarbons like n-octane for separation of non-polar hydrocarbons, glycerol for more polar analytes, and polyethylene glycol.* The last is a synthetic polymeric wax that is available in a number of different molecular weight ranges which, depending upon the range, may be used for the separation of classes of materials of different polarities, like classes of aromatic hydrocarbons or groups of pollutant gases like COS, SO_2, H_2S, CS_2, $(CH_3)_2S$, and methyl and ethyl mercaptans [Mindrup 1978].

Frequently, analytical chromatographers will rely on manufacturers of column packings to suggest appropriate column materials for their applications, and some manufacturers provide proprietary stationary phase formulations in pre-packed columns for specific applications.

In open tube columns, no solid packings are used. Instead, the capillary walls are coated with films of stationary phase liquids, the selection criteria for which are the same as those used for packed columns. But open tube columns provide some significant advantages over packed columns. Because they are coated with very thin films of stationary phase, they tend to have relatively low pressure drops compared to packed columns, which are usually quite tightly packed. Thus, they may be made quite long and still support reasonable, although much lower than packed column, mobile phase flow rates through them. But the thin wall-coating also minimizes the volume of the stationary phase compared to the mobile phase and contributes to reduced retention time (eq. 9.4). So, open tube

*One commercial form of polyethylene glycol is known as Carbowax™, a product of the Union Carbide Chemical Corporation.

columns operate more rapidly, despite their length and their lower mobile phase flow rates. Open tube column length also yields as much as a two order of magnitude increase in the column plate number and a factor of two reduction in the height equivalent of a theoretical plate. Thus, they operate more efficiently as well [Miller 1988, 122].

Detectors

Since virtually any device that is capable of detecting gases and vapors may be used, a large number of general and special purpose detectors are available for application to GC. The two most commonly used, however, are the thermal conductivity (TCD) and flame ionization (FID) detectors.

Thermal Conductivity Detectors

A schematic of thermal conductivity detector is shown in Figure 9.6a. Its operation is based on the measurement of the difference in thermal conductivity between a pure carrier gas and a carrier gas containing an analyte. Typically, a TCD is constructed from a thermostated two chamber device. Mounted in each of the chambers is a resistor with a high temperature-resistance coefficient. Each of the resistors also serves as an arm of a Wheatstone bridge circuit (Figure 8.6b). The resistors are heated and, thus, are maintained at a temperature somewhat higher than the chamber walls. The heating rate applied, however, is very carefully controlled so that the resistor temperatures are dependent only upon the rate at which they can dissipate heat to the walls.

Pure carrier gas is routed through one of the chambers and the chromatographic eluent is routed through the other. Since the temperature of the sensors depend upon their heat dissipation rates, differences in thermal conductivity between the gases passing over them give rise to temperature fluctuations between them, which is sensed by an unbalance in the Wheatstone bridge circuit.

To maximize the sensitivity of a TCD, hydrogen and helium are used as carrier gases because they have thermal conductivities about an order of magnitude greater than most other gases (Table 9.1). Because of the high thermal conductivities of the carrier gas, an analyte gas passing through the eluent side of the detector raises its temperature relative to the reference side.

Despite the large differences in thermal conductivity between hydrogen and helium and typical analyte gases, TCD's are among the lowest in sensitivity of all chromatographic detectors (the applicable range being from about 10^{-8} to 10^{-3} grams). However, they are extraordinarily reliable, relatively inexpensive, and universally applicable to all analytes.

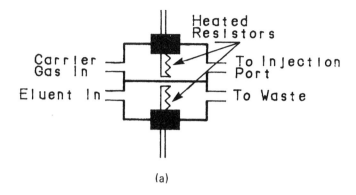

(a)

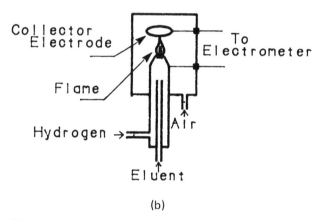

(b)

Figure 9.6   (a) Schematic of a thermal conductivity detector; (b) schematic of a flame ionization detector.

Flame Ionization Detectors

A schematic of a flame ionization detector is shown in Figure 9.6b. Its operation is based on the ionization of analyte molecules when they are burned in a hydrogen flame. The current arising from the transfer of the ions between a collector electrode and the tip of the hydrogen flame is proportional to the amount of analyte being burned in the flame.

The FID is effectively a universal detector for organic compounds, being able to sense any compound with a hydrocarbon backbone. FID's are also considerably more sensitive than TCD's, their range of applicability covering analyte quantities from 10^{-10} to 10^{-4} grams. However, they

Table 9.1   Thermal Conductivities of Some Gases and Their TC Detector
Responses

Compound	Thermal conductivity[a]	Relative detector response[b]
Carrier gases		
Argon	12.5	
Carbon dioxide	12.7	
Helium	100.0	
Hydrogen	128.0	
Nitrogen	18.0	
Typical sample gases		
n-Butane	17.5	85
Cyclohexane	10.1	114
Benzene	9.9	100
Chloroform	6.0	108
Ethane	17.5	51
Ethyl acetate	9.9	111

[a]Relative to helium = 100.
[b]Per mole in helium with benzene = 100.

can not be used to detect any of the small molecule carbon-, nitrogen-,
or sulfur-containing gaseous air pollutants, as Table 9.2 illustrates.

Flame Photometric Detectors

An adaptation of the FID is the flame photometric detector (FPD). In
FPD's, the column eluent is burned in a hydrogen flame and the light
generated is focused on a photomultiplier. Nitrogen is most frequently
used as the carrier gas and is mixed with hydrogen and oxygen at the
column outlet. FPD's may be used for the detection of sulfur compounds,
taking advantage of sulfur's characteristic flame spectrum. Sulfur specificity is achieved by selection of a suitable optical filter.

Table 9.2   Compounds Not Detected by Flame
Ionization Detectors[a]

He	O_2	NO	CS_2	Ar
N_2	NO_2	COS	Kr	CO
N_2O	$SiCl_4$	Ne	CO_2	NH_3
$SiHCl_3$	Xe	H_2O	SO_2	SiF_4

[a]From [Miller 1988, 129].

Photoionization Detectors

Another alternative to the flame ionization detector is the photoionization detector (PID). A PID operates by means of a high energy ultraviolet light source that ionizes analyte molecules as they flow through it. Like the FID, the resulting ion current is proportional to the quantity of analyte passing through. PID's are as much as an order of magnitude more sensitive than FID's. However, early PID designs were usable only at relatively low temperatures and, thus, were limited to detection of volatile organic compounds. Because of their low temperatures, eluting analytes with low vapor pressures tended to condense in them. Newer designs, however, can operate at temperatures as high as 300° C and, thus, maintain even low vapor pressure analytes in the vapor state. As a consequence, the utility of PID's has been extended to include detection of large classes of both aliphatic and aromatic organics, including drugs and pesticides [Langhorst 1981]. PID's may also be used for the detection of many of the organics specified as water pollutants by the EPA [40 CFR, Part 136.3], and they may be used to detect some small molecule pollutants like H_2S and ammonia.

A significant advantage of the PID, in contrast to FID's, is that they are nondestructive, i.e., they do not destroy the analyte molecules as they detect them. Thus, the effluent from a photoionization detector can be routed to a mass spectrometer or to an FTIR for more complete characterization of its components.

Electron Capture Detectors

There are a number of specialty detectors for gas chromatography, perhaps the most common being one that is particularly applicable to detection of picogram to nanogram levels of halogenated and nitrogenated organics as well as to polynuclear aromatics like napthalene and phenanthrene. It is known as an electron capture detector (ECD) and is also an ionization detector. But it works counter to FID's and PID's in that analytes to which it is sensitive reduce instead of increase the ion current. That is, a radioactive source* in the detector ionizes the carrier gas, which for electron capture detection is nitrogen mixed with a small amount of argon and methane, and generates a high base-line ion current. Since the ion current is also dependent upon the concentration of electrons that are produced by the ionization process, when analytes with high electron absorption characteristics, like those mentioned above, enter the detector, they reduce the current in an amount proportional to their quantity.

---

*Radioactive nickel or titanium hydride containing some tritium are ordinarily used.

Coulometric Detection

Although they probably find more applications in liquid chromatography, coulometric detectors, which were discussed earlier in the section on electrochemical methods, are also used for gas chromatography. Sulfur dioxide, for example, may be measured by passing an eluent stream into a coulometric titration cell. Coulometric detection provides an additional level of specificity to the detection of eluent analytes and is excellent for quantitative analysis since the time integral of the current is directly related to the amount of analyte in the eluent stream.

Hyphenated Methods

Although they are not, strictly speaking, chromatographic detectors, mass spectrometers and Fourier transform infrared spectrometers are finding increasing application as analyzers of chromatographic effluents. As has been mentioned, when used as such, reference is usually made to hyphenated methods, i.e., GC-Mass Spectrometry or GC-FTIR.*

The most successful of the hyphenated techniques has been GC-MS. EPA, in fact, prescribes GC-MS as the method of choice to confirm results that are obtained by other methods when unfamiliar pollutants are analyzed [40 CFR Part 136.3]. However, GC-FTIR has also seen increasing success recently. Thus, a large number of commercial instrument packages incorporating one or both of the methods into chromatographic systems are now available.

Apart from their measurement principles, an important distinction between GC-MS and GC-FTIR is the way in which the chromatographic column is interfaced to one or the other analyzers. Recall that the ionization process in a mass spectrometer is carried out under high vacuum conditions. Mass spectrometry, therefore, requires that the eluent from the column be reduced in pressure before its entry into the spectrometer.

For GC-MS, capillary chromatography seems to be the preferred separation method because the flow rates used are low enough such that direct entry into the spectrometer is possible, the pumping speed of the spectrometers used apparently having sufficient capacity to maintain appropriate vacuum conditions. When packed columns are used, however,

---

*It must be noted here that hyphenated techniques are not for the uninitiated. Specifically, the EPA requires that technicians using GC-MS, for example, be able to demonstrate acceptable precision and accuracy when analyzing test samples. EPA also demands on-going quality control programs in laboratories using GC-MS. And finally, hyphenated methods should be regarded to be of a level of sophistication requiring experienced specialists to oversee their proper use.

the eluent stream may not be injected directly into the spectrometer. It must be passed through a porous permeation tube or by a porous membrane to control the entry rate of the eluent molecules.

In contrast to GC-MS, large amounts of analyte are beneficial to analysis by FTIR. Thus, in GC-FTIR, the column eluents are routed directly through a heated, long path, infrared cell, usually made of glass and coated with an IR reflecting gold film on its interior surface.

GC-MS and GC-FTIR are about equal in sensitivity. They are also complementary methods. That is, while mass spectrometry cannot distinguish isomers,* it can provide information to establish empirical chemical formulas through analysis of mass numbers. FTIR analysis, on the other hand, cannot be used to determine molecular formulas but it can distinguish among functional groups and so, suggest the structure of the material being analyzed. FTIR is also non-destructive. Thus, a very powerful analytical tool resides in the combination of MS and FTIR with chromatography. Passing the eluent first through a non-destructive detector like a TCD or a PID, provides preliminary identification via the unknown's retention time and quantitation via its peak height or area. Passing it then through an FTIR followed by MS can provide unequivocal identification.

LIQUID CHROMATOGRAPHY

Equipment in General

Precursors to modern liquid chromatographs were relatively simple, consisting of a solvent reservoir, a wide bore glass column, particle packings of relatively large diameter, and a series of vessels (a fraction collector) to collect eluent as a function of time. Mobile phase was passed through the column under the influence of gravity alone. Analysis followed elution by applying appropriate physical or chemical methods to the eluents in the fraction collector.† Modern analytical systems, however, in order to increase separation speed and enhance efficiency, make use of high pressures so that the mobile phase may be forced through much narrower columns containing much denser, finer packings at rates comparable to those used for gas chromatography. And, like gas chromatography, detection is usually via an on-line device that continually monitors the eluent.

---

*Isomers are chemical compounds that have the same elemental composition but different structures.

†This older method is still used in many bioanalytical methods as well as in preparative chromatography, in which the separation capabilities of the method are used to isolate desirable products from complex biological mixtures.

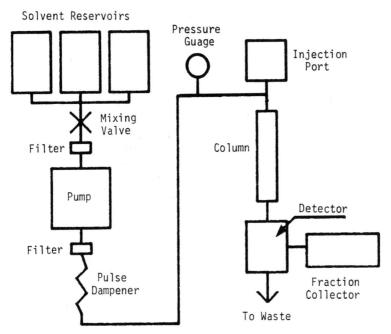

Figure 9.7   The components of a typical liquid chromatograph. Multiple solvent reservoirs may be used for gradient elution chromatography by appropriately programming the mixing valve (see text).

A schematic of a liquid chromatograph is illustrated in Figure 9.7. The essentials of the instrument include a solvent reservoir, a high pressure pump, filters to prevent extraneous particulates from entering the column or the pump, a flow controller, a sample injection port, the column, and a detector.

Pumps

Several types of pumps may be used, each of which provides its own set of advantages and disadvantages. Syringe pumps, which are essentially motorized syringes, provide pulse free flow at pressures up to about 7500 lb/in^2. Their primary disadvantage is that they provide only a fixed volume of mobile phase.

Reciprocating piston and diaphragm pumps are probably the most commonly used in liquid chromatography. They are capable of providing continuous and precisely controlled mobile phase flow rates, but their reciprocating action imparts a pulsation to the mobile phase flow that results in a fluctuating base line in the detector. Thus, reciprocating pumps require some kind of pulse dampening device in the circuit.

Constant pressure pumps, which make use of some form of pneumatic device to directly pressurize the solvent reservoir, provide reliable pulse-free flows and are relatively low cost and simple. However, they are not capable of providing flow rates with the precision of reciprocating devices.

Injection Ports

Injection ports are similar in principle to those used in gas chromatography, with injection directly into the column providing the best separation efficiency. Thus, syringe injection systems have been designed that can operate into pressures as high as 1500 lbs/in^2. A schematic of such a port is shown in Figure 9.8. In very high pressure LC systems, however, slide valves similar to the one shown in Figure 9.5 must be used since they are capable of injecting sample into columns being run under pressures as high as 7000 lbs/in^2.

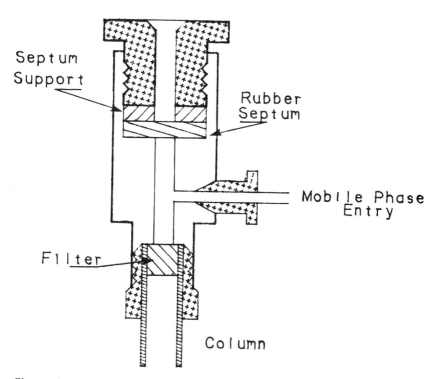

Figure 9.8   A liquid chromatographic syringe injection port.

Isocratic and Gradient Elution

As was mentioned earlier, gas chromatography may be carried out iso-thermally or with temperature programming to reduce the retention times of analytes with high valued partition coefficients. Temperature is less effective in enhancing separations in liquid chromatography. However, liquid chromatography has its analogous operational modes. That is, since both the stationary and mobile phases contribute to separation in LC, the solvent character of the mobile phase may be used as an additional adjustable parameter to enhance separation. Thus, one may use either a mobile phase of fixed chemical composition or change its composition in some systematic way throughout the course of a run. Use of constant composition is called *isocratic elution* while a programmed change in composition is called *gradient elution*. Gradient elution requires the use of two or more solvent reservoirs (Figure 9.7), which supply the components required to vary the mobile phase composition. The parameters that may be thus adjusted include pH, ionic strength, or solvent polarity.

Columns

For most analytical applications, columns are straight sections of stainless steel tubing of lengths varying between 10 and 100 cm with diameters of about 5 mm. Although coiled and "U" shaped columns may be used, they tend to be less efficient than straight columns unless the ratio of the radius of curvature of the coil to the length of the column is greater than 130 [Hamilton & Sewell 1982, 58]. Most commercial columns are straight and about 25 cm long. The particle size of typical packing materials ranges between 5 and 50 microns. Such columns produce plate numbers with values as high as 25,000.

Detectors

The range of detection methods used in liquid chromatography parallels the breadth of those used in GC. They include photometric detectors based on ultraviolet and infrared absorption and fluorescence, electro-chemical detectors based on amperometry and conductometry, adaptations of GC methods like PID's, FID's, ECD's, mass spectrometry, and detectors based on the difference between the refractive index of pure mobile phase and mobile phase carrying an analyte band.

With the exception of the last, the operational principles of all of the methods have already been discussed. But the last is second only to ultraviolet absorption in popularity as an LC detector because it has, for all practical purposes, universal applicability, analogous in that sense to thermal conductivity in GC. This is in contrast to uv absorption which, of course, requires that the eluting analytes have the appropriate light absorption characteristics. Thus, we shall present some of the basic principles of refractometry as used in LC detection.

Refractometry

Refractometry is based on the change in the velocity of light waves as they pass between transparent media with differing dielectric properties. As was discussed in the section on light dispersion devices, if the angle of incidence of the source light is not normal to the plane of the interface between the two media, the light is deflected to a new angle as it passes through the second medium (Figure 7.5a). The change in direction of the light is called refraction and the relationship between the angles of incidence and refraction is given by Snell's law

$$n_1 \sin \phi_1 = n_2 \sin \phi_2$$

where n_1 and n_2 are the refractive indices of media 1 and 2, respectively, and ϕ_1 and ϕ_2 are the angles of incidence and refraction, respectively.

Most commercial refractive index detectors are based on one of two principles, the deflection of a light beam or Fresnel reflection. The first, the deflection detector, depends directly upon the change of refraction angle resulting from a change in refractive index in the chromatographic eluent. Fresnel detectors are based on a change in the ratio of reflected to transmitted light that occurs as it crosses an interface between two media, when the refractive index of one of them changes.

Deflection Detectors

In a deflection detector, the eluent is routed through a cell constructed in the shape of a hollow prism. Light incident upon the prism is then refracted to an extent determined by the eluent's refractive index. If the light exiting the prism is initially focused on a phototube when pure mobile phase is passed through it, it will be deflected away from the phototube as analyte bands are carried through the cell because of changes in refractive index, and the deflection will generate a signal fluctuation. This is the basic principle of all deflection-type refractive index detectors. However, the optical sophistication of commercial detectors is considerably greater than the simplified prism cell we have described.

Fresnel Detectors

The Fresnel detector is based on Fresnel's law of refraction and reflection, which states that part of the light incident upon an interface between two different transparent dielectric media will be reflected from the interface and part will pass into the second medium and be refracted. The formalism of Fresnel's law* is

$$R = \frac{1}{2}\left[\frac{\sin^2(\phi_1 - \phi_2)}{\sin^2(\phi_1 - \phi_2)} + \frac{\tan^2(\phi_1 - \phi_2)}{\tan^2(\phi_1 + \phi_2)}\right]$$

where R is the ratio of reflected light intensity to that of the incident light and ϕ_1 and ϕ_2 are the angles of incidence and refraction, respectively. Now since the indices of refraction are related to the angles of incidence and refraction through Snell's law, it can be seen that a change in refractive index in one of the media will lead to changes in the ratio of reflected (and transmitted) to incident light.

A diagram of a simple Fresnel detector is shown in Figure 9.9 [Conlon 1961]. The device is constructed from a bent light pipe that transmits source light to an optically flat window sealed into a glass tube through which eluent from a chromatograph is routed. The bend angle of the light pipe allows a fraction of the light from the source to be reflected to

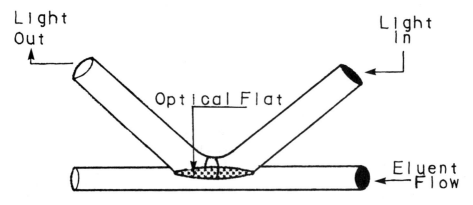

Figure 9.9 A simplified Fresnel detector. Light is partly reflected from the optical flat to the exit of the light pipe and partly refracted through it into the eluent flow tube. A change in refractive index in the eluent changes the intensity of the reflected portion (see text).

---

*Fresnel's law is derived by consideration of energy conservation as light is reflected and transmitted through an interface between two transparent dielectric media [Kortum 1969, 5].

a photo cell and a fraction to be transmitted through and refracted by
the flowing mobile phase. Changes in the composition of the mobile
phase then, as it elutes analytes from the chromatograph, change its
refractive index and, concurrently, the amount of light reflected to the
phototube. Like the deflection detector, the optical sophistication of modern Fresnel devices is very much greater than the simple Conlon cell
described, but they function is essentially similar ways.

Stationary Phases, Mobile Phases, and Catagories of LC

The range of operating modes in LC is much larger than in GC because
both the stationary and the mobile phases influence separation. That is,
one may take advantage of solvent characteristics in both phases to effect
separation.

Standard and Reversed Phase Liquid Chromatography

In general, the name given to the specific application of liquid chromatography is determined by the nature of the mobile and stationary phases
that are used. Thus, in traditional or *normal phase* LC, the stationary
phase is made the more polar material. However, many modern systems
make use of polar mobile phases and non-polar stationary phases. When
the mobile phase is the more polar, the operational mode is called *reversed
phase* chromatography.

Normal and reversed phase partition chromatography involve the use
of stationary phases composed of a liquid film coated on or chemically
bonded to an inert substrate and a mobile phase of opposite polarity.
Chemical bonding precludes the possibility of the liquid film being "stripped" from the substrate by the mobile phase. Thus, many stationary
phases are prepared from silica, the surface of which has been modified by
reacting it with a class of compounds called organosilanes. The resulting
chemically reactive surface is of the general form

$$SI-O-Si-R$$

where R may be selected to satisfy the specific requirements of either
normal or reversed phase separations.

In normal phase liquid-solid *adsorption chromatography*, a polar mobile phase like water or ethylene glycol may be used with a non-polar
stationary phase like finely divided polyethylene or polypropylene pow-

ders. In reverse phase adsorption chromatography, an aliphatic hydrocarbon may be used for the mobile phase in combination with a polar stationary phase like silica, glass, or alumina.

Ion Chromatography

Ion exchange or *ion chromatography* is a form of liquid-solid chromatography that has specific application to the separation of ionic species dissolved in an aqueous mobile phase. Separation is the result of ion exchange processes that occur between the mobile and stationary phases, the latter of which are prepared from a class of materials called ion exchange resins.

Ion exchange resins are polymeric materials with ionic functional groups attached to them. Typical resins have the structures

$$\left[\begin{array}{c} | \\ -R'-SO_3^-H^+ \\ | \end{array} \right]_n \quad \text{and} \quad \left[\begin{array}{c} | \\ -R'-NR_3^+Cl^- \\ | \end{array} \right]_n$$

where R' is usually an aromatic group and R is usually a methyl or ethyl group.* The structure on the left is that of a *cation exchange* resin in that positive ions in the mobile phase may exchange with the hydrogen ions on the resin. The one on the right is that of an anion exchange resin and anions in the mobile phase may exchange with its chloride ions.

Ion exchange involves an equilibrium that is established between the mobile phase and the resin according to

$$R^-A^+ + B^+ = R^-B^+ + A^+$$

with the equilibrium for the reaction determining the distribution of analyte ions between the two phases. Thus, for example, the initial state of a cationic exchanger might be all H^+ form. Cationic analytes injected into the column then bind to the resin, displacing the hydrogen ions. But the pure mobile phase, carrying its own high concentration of cations, then elutes them back off and carries them through the system.

Amperometric, coulometric, and conductometric detectors are all used in ion chromatography. Conductivity, however, has become the more commonly used detection method since the advent of so-called *suppressor columns* to treat the eluent before it enters the conductivity detector.

---

*The structure $-NR_3^+$ is called a quaternary ammonium group.

Requirements for suppression arise because of the intrinsic conductivity of the mobile phases used in ion chromatography. The function of the suppressor column is to render the mobile phase non-conductive and is best explained by the example shown in Figure 9.10,* which illustrates a column configuration that might be used to analyze a sample of acid rain. For an ion mixture of the kind shown, the stationary phase is an anion resin in the bicarbonate form, and the eluent is a solution of sodium bicarbonate, a good ionic conductor. The initial separation processes occur according to

$$\text{Resin}-\text{N}^+\text{HCO}_3^- + \text{Na}^+\text{X}^- = \text{Resin}-\text{N}^+\text{X}^- + \text{Na}^+\text{HCO}_3^-$$

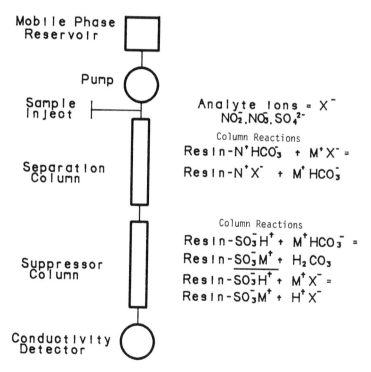

Figure 9.10 A schematic of the reactions underway in an ion chromatographic system in which conductometric detection and a suppressor column are used.

*For a full discussion of suppressor column chemistry, see [Wetzel et al, 1985] or [Fitchett, 1983].

where X^- refers to the anions in the mixture, which are bound initially to the stationary phase.

One might expect that it would be possible to detect a change in conductivity of the column eluent at this point because of the increase in bicarbonate concentration resulting from the exchange. However, the conductivity of the $NaHCO_3$ of the mobile phase is very high relative to any change it might experience as a result of the exchange reaction. Thus, the binding reaction contributes little to conductivity changes in the eluent.

Continued flow of fresh mobile phase reverses the binding reaction and frees the ions, with retention times related to their equilibria with the resin. With the displacement of the analyte ions from the column, the eluent becomes a mixture of sodium bicarbonate and separated zones of the sodium salts of the analyte ions. These, upon entering the suppressor column, encounter an exchange resin in acid form; the sodium bicarbonate is changed to carbonic acid and the sodium salts of the analyte ions become strong acids of the form H^+X^-. Thus, the strongly conducting sodium bicarbonate eluent is converted to weakly conducting carbonic acid, and the intermediately conducting sodium salts are converted to strongly conducting mineral acids (see section on equivalent conductivity). The post-treated effluent, therefore, exhibits low conductivity except when it contains a band of analyte ions.

Suppressor columns, thus, serve a dual function. They suppress the conductivity of the pure mobile phase and, at the same time, by converting them to their mineral acid analogs, enhance the conductivity of the analyte ions.

More modern adaptations of suppressor columns make use of ion selective membranes instead of packed ion exchange resins. These newer suppressor devices eliminate the need to regenerate the ion exchange resins used in the columns, a major disadvantage that they exhibit. That is, as the mobile phase passes through the resin packed suppressor, the resin is progressively saturated with the flowing eluent cation. Thus, the resin must be converted back to its protonated form periodically to maintain its efficacy. Ion selective membrane suppressors preclude the necessity to regenerate resins.

Ion selective membranes are, essentially, ion exchange resins that are cast in the form of films or hollow tubes. Their utility lies in their ability to allow only ions of the same charge as their exchange moieties to permeate through them. That is, cation selective membranes (made from cation exchange resins) permit only cations to permeate through them, and anion selective membranes (made from anion exchange resins) permit only anions to permeate through them. Thus, in the newer suppressor devices, the eluent containing the analyte ions from the separation column is passed through the interior of the ion selective membrane tube or between ion selective membrane films. The outsides of the tube or film devices are, in turn, bathed by a counter-current flow of strong acid in the case of cation suppressors or strong base in the case of anion suppressors. For example, in the case of a cation suppressor, protons from the flow of sulfuric acid can permeate a cation selective membrane suppressor and exchange with the flowing cations of the mobile phase as well as with the separated analyte cations in the interior. In turn, the analyte and eluent cations permeate back out through the membrane where they are carried away by the counter-current flow of acid bathing the exterior of the membranes. The net effect is identical to that of the resin packed columns, i.e., the eluent is converted to a low conductivity weak acid, and the analyte ions are converted to their strongly conducting mineral acid counterparts.

Exclusion Chromatography

Gel permeation chromatography and gel filtration are names given to the liquid-solid form of exclusion chromatography. As was mentioned earlier, the stationary phase used in exclusion chromatography is made from polymer gels with well defined pore sizes and distributions, a commercial example of which is a dextran material marketed under the name Sephadex™.

Exclusion chromatography separates substances according their molecular sizes and shapes. Small molecules may freely enter the pores of the gel while very large molecules are excluded. Thus, small molecules are the most retarded in their transit through the system. and their partition coefficients are defined as equal to unity.* On the other hand, very large molecules are completely excluded and are the most rapidly eluted; they are considered to have partition coefficients equal to zero. Intermediate sized molecules exhibit partition coefficients between zero and one and retention times that are inversely proportional to their molecular size.

THIN LAYER CHROMATOGRAPHY

Thin layer chromatography, as mentioned earlier, is a form of liquid chromatography carried out on a stationary phase immobilized on a plane surface. Separation is accomplished by allowing the mobile phase to

---

*In contrast to standard chromatography, in which the mobile phase is the most rapidly transported component of the eluent. in exclusion chromatography. it is the most retarded, because mobile phase molecules are small.

migrate across the plane surface by capillary action, carrying analyte molecules along in much the same way as in column methods.

The stationary phase may be of any material that can be finely divided and formed into a uniform layer, but the most popular material is silica gel. Plates are prepared by slurrying the stationary phase material with a binder like polyvinyl alcohol in water and spreading the slurry onto a plate about 20 cm on an edge and in thicknesses of the order of 0.25 to 0.5 mm. Although preparing TLC plates is not difficult, the easy availability of commercial plates makes purchasing them the preferred direction for most laboratories.

Selection of mobile phases for TLC is much the same as for column techniques. Since silica gel represents a polar stationary phase, non-polar mobile phases are used with it, occassionally modifying them by mixing with more polar solvents. The flow of the mobile phase is dependent upon its viscosity and its surface tension with respect to the stationary bed as well as on the physical configuration of the bed with regard particle size, particle density, etc.

Analysis by TLC is carried out by spotting small samples of analyte about one cm inside one edge of the plate with a micropipette. The plate is then placed in a "development" chamber (Figure 9.11) with the spotted

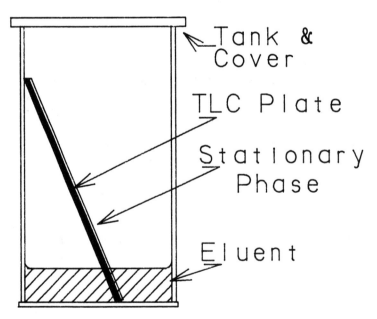

Tank &
Cover

TLC Plate

Stationary
Phase

Eluent

Figure 9.11   A development tank for development of thin layer plates.

edge immersed in the mobile phase to a level just short of immersing the sample spots. As in column chromatography, the mobile phase leads the analyte molecules, separating analyte bands along its transit path. When the mobile phase front has nearly reached the opposite edge of the plate, the plate is removed from the chamber and dried.*

Analyte detection may be based on the natural color of the analyte molecules or reactive dyes may be sprayed onto the plate to color the bands. Detection may also be based upon the fluorescent properties of the analytes, which may be stimulated by shining uv light on the plate. In some applications, the stationary phase is coated with a phosphorescent material, with analyte molecules being detected by extent to which they quench the phosphorescence. The appearance of a thin layer plate as a function of time is illustrated in Figure 9.12.

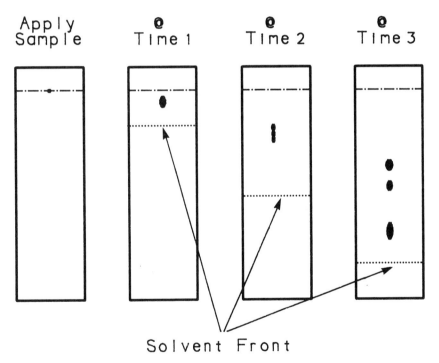

Figure 9.12   A thin layer plate as it might appear if analyte bands were visualized as a function of time.

---

*Two dimensional TLC may also be carried out by re-chromatographing the plate after rotating it 90° from its original orientation.

Analyte identification is made through measurement of the position of the bands relative to the distance traveled by the mobile phase front, determining for each analyte an R_f factor

$$R_f = \frac{\text{distance traveled by band}}{\text{distance traveled by mobile phase}}$$

which is analogous to the retardation factor of column chromatography.

Quantitative analysis is carried out by measuring the intensity of the developed analyte band, for example, the optical density of a colored band or the flourescent intensity of a fluorophore.

ENVIRONMENTAL APPLICATIONS OF CHROMATOGRAPHY

Gas Chromatography in Air Pollution

Although gas chromatographic methods have been successfully used for the separation and quantitation of the permanent gases of the air* as well as for carbon, nitrogen, and sulfur oxides and H_2S, the method is not among the reference procedures specified by the EPA for the analysis of the criteria air pollutants — except for hydrocarbons. But even though hydrocarbon pollutants may be separated on open tube capillary columns and identified via GC-MS techniques or by means of FIDs and PIDs, detailed analyses of them in air appears to be limited to research studies, perhaps because of the enormous number of hydrocarbons that are always present in air as a result of both natural and anthropogenic processes [Braman 1983, 136]. Regulated testing is directed primarily to determination of total hydrocarbon values and the level of only one specific contaminant — methane.

The EPA reference method [40 CFR Part 50, Appendix E] for the determination of total hydrocarbons and methane is based on first measuring the total value by passing the sample directly through a flame ionization detector. A second sample is then stripped of water, carbon dioxide, and all hydrocarbons other than methane by passing it through a cryogenic column where all but carbon monoxide and methane are condensed out of it. The non-condensed gases are then passed into a chromatographic column where the methane and CO are separated. Measurement of CO and methane is also with a flame ionization detector. But since FIDs are not sensitive to CO, before the gases are measured,

---

*Oxygen, nitrogen, argon, water, and carbon dioxide.

they are mixed with hydrogen and routed through a catalytic converter where the CO is reduced to methane. Hydrocarbon concentrations corrected for methane are then determined by subtraction.

Other typical environmental applications for gas chromatography include determination of levels of polycyclic (also called polynuclear) aromatic hydrocarbons (PAHs and PNAs) in the air. This family of compounds is the subject of considerable study because of the carcinogenic character of its members. Capillary gas chromatography combined with PIDs and FIDs may be used to characterize the complex mixtures of these compounds, which are generated by automotive exhaust, coal-fired power plants, and cigarette smoke, and are found as components of pollutant particulates.

Halogenated hydrocarbons, including the chlorofluorocarbons responsible for depletion of the stratospheric ozone layer, may also be separated and quantitated chromatographically. The high sensitivity of the electron capture detectors used for halogenated hydrocarbons permits detection to ng levels for many species.

Peroxyacetylnitrate (PAN), hydroxyperoxides, and alkylperoxynitrates, the precursors and components of photochemical smog, are also determined in air by means of electron capture detectors after separation on Carbowax^{TM} stationary phases.

Liquid Chromatography in Air Pollution

The primary air pollution application of liquid chromatography has been the use of normal and reversed phase methods to characterize PAHs. Liquid chromatography, in fact, provides some advantages over capillary column gas methods because some enhancement in sensitivity is obtained by using ultraviolet and fluorescent detection. In reverse phase methods, column eluents have included such non-polar solvents as cyclohexane, benzene, and petroleum ether in combination with stationary phases of silica gel or alumina, while normal phase methods have made use of methanol-water eluents in combination with stationary phases of long chain hydrocarbons bound to silica or alumina [O'Brien 1983, 160].

Thin Layer Chromatography in Air Pollution

One of the more common applications of TLC in air pollution studies is its use as a pre-separation technique for gas or liquid chromatographic analyses of PAHs [Zelinski and Hunt 1983, 177]. Pollutant particulates from the air are first collected on glass fiber filters. The filters are then washed with solvents that extract the PAHs, and the extracts are applied to TLC plates. After thin layer separation, the analytes are individually extracted from the TLC plates for final gas or liquid chrmomatographic characterization.

Gas Chromatography in Water Pollution

Although all three chromatographic methods, GC, LC, and TLC may be applied to analysis of water pollutants, gas chromatography predominates in actual use. In fact, EPA prescribes specific chromatographic protocols for the determination of nearly 200 different regulated pesticide and non-pesticide organic compounds [CFR Part 136, Appendix A], and the methods prescribed in all but a handful of them involve gas chromatography. The pollutants represented in the regulated compounds include detergents, plasticizers, pesticides, products of water chlorination, industrial wastes, and the decomposition products of many of them. Table 9.3 is a summary of the EPA methods used for determination of several chemical classes that represent many of these compounds.

In general, organic contaminants must first be separated from polluted waters before they can be introduced into a chromataograph. Furthermore, levels of water pollutants are often very low (of the order of μg per liter or parts per billion) and their analysis frequently requires that they be preconcentrated. Thus, because of the large numbers of organics that polluted waters may contain, they have been classified into three general groups, each reflecting the methods by which they are separated from water or concentrated for subsequent analysis [Walton 1983 264].

The categories include *purgeables*, *extractables*, and *intractables*. The purgeables are compounds of low molecular weight that lend themselves to concentration by purging, i.e., by passing an inert gas through the water sample and concentrating the analyte vapors by carrying them through a cold chamber to freeze them or adsorbing them on a high surface substrate. Purge and trap input devices, clearly, have wide application for analysis of such volatile organic pollutants.

Extractables are compounds that may be extracted into and concentrated in a suitable organic solvent by shaking it with the water sample. Extractables are usually sufficiently volatile to introduce directly into a gas chromatograph once they have been removed from the water and concentrated.

Intractables are compounds that must be separated from water by sorption onto activated carbon or similar high surface materials because they are neither volatile nor do they easily lend themselves to extraction. They include such compounds as polar organics, carbohydrates, and organic acids. Sorbed intractables, thus concentrated, may be analyzed by subsequent extraction from the sorbent.

Table 9.3   EPA-Prescribed Methods for Analysis of Organics and Pesticides in Water

EPA No.	Method type	Detector	Detection limit $\mu g/l$	Analyte class
601	GC + purge and trap	ECD	4–10	Halocarbons
602	GC + purge and trap	PID	0.2+0.4	Purgeable aromatics
603	GC + purge and trap	FID	0.5–0.7	Acrolein and acrylonitrile
604	GC[a]	FID or ECD[b]	0.3–20[c]	Phenols
605	LC[d]	Amperometric	0.08–0.13	Benzidines
607	GC[a]	ECD	0.2–3	Pthalate esters
608	GC[a]	ECD	0.003–0.25	Organohalide pesticides and PCBs
609	GC[a]	FID and ECD	3.0–20.0	Nitro-aromatics
610	GC[a] or reverse phase LC[a]	FID for GC uv and fluorescence	0.02–0.04	PAH's
611	GC[a]	ECD	0.3–4	Haloethers
612	GC[a]	ECD	0.05–1	Chlorinated hydrocarbons
613	GC	MS	0.002	Dioxin
624	GC and purge and trap	MS	1–10	Purgeables
625[e]	GC[f]	MS	1–20	Acid and base extractables

[a]Methylene chloride extraction.

[b]Confirming method.

[c]Depends on analyte.

[d]Methylene chloride extraction followed by exchange with alcohol and addition of buffer solution.

[e]GC-MS method 625 is recommended by EPA to be used as a confirmatory procedure for unfamiliar unknowns of all types.

[f]Methylene chloride extractions from samples that are pre-treated to make them acidic, neutral, or basic, each of the conditions allowing for a different group of analytes to be isolated.

Some organic pollutants, particularly the intractables, require chemical modification to make them more amenable to separation from water or to chromatographic analysis. Such modifications are called *derivatizations*. Examples of derivatizations include conversions of organic acids or phenols into methyl esters to enhance their volatility, conversion of some herbicides to their trifluoroacetyl derivatives to enhance their chromatographic separability, and the formation of halide derivatives of the pollutants to achieve the enhanced sensitivity of electron capture detectors [Guiliany 1983].

Liquid Chromatography in Water Pollution

Although liquid chromatography provides several advantages for analysis of water pollutants, particularly for intractables in waste water, its primary use has been for collection and concentration [Walton 1983]. That is, eluent from an LC column is collected in a fraction collector. The fractions thus collected are then analyzed by gas chromatographic methods.

The chief disadvantage of LC for water pollution is that it provides too much information. Typical high resolution chromatograms from waste waters may contain too many peaks to provide reliable identifications, even when analyte-specific detection methods like fluorescence or electrochemical activity are used. Although there has been some progress in coupling mass spectrometers to liquid chromatographs, there is an intrinsic incompatability between the two methods that has limited successful application of that hyphenated technique. That is, the liquid eluent from the LC must somehow be converted to a gas to be injected into the high vacuum chamber of the mass spectrometer. It is for this reason that the method of choice for many pollutants is GC-MS on analytes first separated and concentrated by LC techniques.

Thin Layer Chromatography in Water Pollution

Although the EPA prescribes TLC for the determination of several organic phosphate insecticides [40 CFR Part 136.3], TLC, like LC, is frequently used as a method to separate and concentrate analytes for subsequent gas chromatographic analysis. Bands on the TLC plate may be extracted with a suitable solvent or the band may actually be cut or scraped from the plate and processed for subsequent analysis. Additionally, because it is inexpensive and its protocols are simple, TLC is frequently used in LC method development, because it provides a convenient way to select optimum mobile and stationary phases for specific analytes.

TLC is also used as a screening technique, particularly in field studies, where its simplicity allows for rapid qualitative analysis of contaminants in surface waters and bottom sediments, and it may be used as a screening technique leading to subsequent analysis by gas chromatography of PAH's, herbicides, and pesticides [Hunt 1983, 302].

Acid Rain and Ion Chromatography

Ion chromatography is the EPA specified procedure for the determination of sulfates and nitrates in acid rain [Operations and Maintenance Manual, Part II, Section 4.8]. The prescribed column components include an ion exchange resin with quaternary ammonium exchange groups as the stationary phase and a mobile phase of a solution of sodium carbonate and sodium bicarbonate. Detection is via electrical conductivity after passing the eluent through a strong acid suppressor column that converts the analyte ions to their acid forms and the mobile phase to carbonic acid.

10

Chemical Methods

INTRODUCTION

Earlier, we defined chemical methods as those that make use of chemical transformations as the basis of separation and quantitation. One either measures the amount of product generated by the transformation or the amount of reactant necessary to achieve the transformation and then relates those values to the amount of analyte in the unknown sample. Such methods have served as the cornerstone of chemical analysis since the beginning of the study of modern chemistry. Thus, for example, Antoine Lavoisier, after learning from Joseph Priestly in 1774 how to prepare oxygen, determined the elemental compositions of combustible organics by burning them in oxygen and collecting and weighing the water and carbon dioxide combustion products. About 50 years later, a German chemist named Justus Liebig learned that better quantitation could be obtained by burning the organics in the presence of copper oxide, which served as an additional source of oxygen. And at about the same time, Jean Baptiste Dumas, a French chemist, developed a method to determine nitrogen in organic matter by burning it in the presence of copper oxide to form nitric oxides and carbon dioxide. The product gases from the combustion were then passed over metallic copper where the oxides of nitrogen were reduced to elemental nitrogen, which was then separated from the carbon dioxide by absorbing the latter in a solution of potassium hydroxide.

Interestingly, despite the vast array of analytical instruments available to the modern analyst, only some of which have we described in the foregoing discussions, these old classical chemical transformation methods of analysis remain a mainstay of modern analytical chemistry. Methods have been streamlined and manipulations have been simplified and many of them automated. But the principles of many of the methods remain essentially identical to those first developed by very early investigators.

There is an extraordinary diversity of chemical methods used in analytical chemistry. The following discussion, therefore, is limited to several categories that are frequently applied to determination of our subject elements in the environment. These include gravimetric methods in which the analyte's chemical transform is simply weighed, volumetric methods in which the volume of a reactant used to quantitatively transform the analyte into an identifiable product is measured, and combustion methods in which the analysate is burned in an oxygen environment and the products of the combustion are related to the analyte quantity.

Continuous flow analyzers are also discussed as examples of automated environmental analysis. Continuous flow analyzers are devices in which the analysate is combined with or injected into a flowing stream of transformation reactants. The chemical transformation occurs as the stream moves through the analyzer, and the products are appropriately measured just previous to the stream's exit from the system.

Finally, a brief discussion of gas detection tubes, devices that may be used for rapid estimation of levels of toxic gases and vapors in air, is also presented.

GRAVIMETRIC ANALYSIS

Gravimetric analysis requires that the constituent being determined be separated from the sample. This may be accomplished by precipitating a chemical transform of the analyte and separating, drying, and weighing the precipitate. Alternatively, if the analyte can be transformed into a volatile product, it may be separated from the analysate by purge techniques or evaporation and collected on an adsorbent and weighed, or the loss in weight of the residual analysate may be determined.

Gravimetric precipitation analysis may only be conducted when the solubility of the precipitant in the reaction medium is negligible, when the precipitate is readily separable from the medium, and when its composition is truly representative of the analyte's composition. Thus, a precipitate must be filterable or of such density that it may be separated by centrifugation, and it must be in a state of subdivision that minimizes the possibility of its dragging foreign substances with it by occlusion or adsorption.

Gravimetric volatilization methods require either that a suitable absorption medium be available for the volatile constituent or that the residual component be sufficiently stable to use its weight loss as the measure of analyte quantity. An example of a need to have a suitable absorption medium might be for the determination of the water content of a carbonate salt. Heating would both drive off the water and decompose the salt. Thus, the residual weight would not reflect the water content of the sample, and the evolved vapor would be a mixture of carbon dioxide with the water analyte. A suitable dessicant would be required that would absorb the water but not the CO_2.

Conversely, one might determine the amount of carbonate salt in a wet sample by heating and decomposing it to the oxide and carbon dioxide. Again, in order to properly measure the quantity of the evolving CO_2, it would be necessary to separate the evolving water vapor and CO_2 with a dessicant. Separation of these gases from one another is an important technology since the evolution of water and carbon dioxide from burned organic compounds serves as the basis of their elemental analysis. Appropriate absorption media will be described in the section on combustion methods.

Advantages and Disadvantages of Gravimetric Methods

Among the advantages of gravimetric analyses are high sensitivity and low cost. Equipment requirements are minimal, and the methods can usually be carried out by relatively unsophisticated laboratory personnel. Sensitivity can be made as high as necessary by simply using larger samples so as to yield higher product weights. The prime disadvantage of the methods are time requirements and labor intensity. That is, gravimetric methods can frequently be protracted because of complicated manipulations required to separate and purify the reaction product. And gravimetric methods do not readily lend themselves to automation.*

Applications of Gravimetric Methods

A major application for classical gravimetric analysis in environmental monitoring is for the determination of sulfate in waste and drinking waters. The gravimetric method is one of several EPA reference methods for sulfate, the details of which are presented in ASTM Method D 51682A [ASTM Annual Book 1987]. The method consists, basically, of precipitating highly insoluble barium sulfate from the water sample by treating it

---

*However, with the advent of laboratory robots, many gravimetric analyses may now be effectively automated [Laboratory Robotics Handbook 1988].

with a solution of barium chloride. Although the method is claimed to be sensitive to levels of sulfate of the order of several mg/liter (PPM), barium sulfate precipitates are notorious for their ability to occlude or adsorb co-species from typical analysates [Kolthoff et al. 1969, 603]. Thus, the alternative EPA methods are probably preferable for the most accurate analyses.

VOLUMETRIC ANALYSIS

Volumetric analysis or titrimetry, as it is frequently called, is based on the addition to an analysate of a reagent that is known to react with the analyte. By using a known concentration of *titrant* and adding enough of it to achieve chemical equivalence with the analyte, one can relate the added volume to the analyte concentration in the analysate. Titrimetric methods used in environmental analysis include acid-base titrations and oxidation-reduction or *redox* titrations.

In both acid-base and redox titrations, the point at which chemical equivalence occurs may be determined in a number of ways, the choice of which depends upon the system being analyzed. Thus, both methods may make use of indicator dyes that change color as soon as a slight excess of titrant is added to the analysate. Such color changes can be detected either visually or by means of a spectrophotometer. Alternatively, one can exploit the electrochemical properties of the analyte or the titrant to detect equivalence points. Thus, one can measure the potential* or the amperometric current changes in oxidation-reduction systems, or the conductivity of systems in which the titrant-analyte reaction product is less conductive than either by itself (see Chapter 8, "Electrochemical Methods").

Acid-Base Titrations

In acid-base titrations, an acid titrant is used to determine a basic substance or a basic titrant is used to determine an acid substance. The former is sometimes called acidimetry while the latter is called alkalimetry.

Definitions of Acids and Bases

But what are acids and bases? Although there are alternative definitions for them, the one most useful for chemical analysis defines an acid as a substance that tends to donate protons to other substances and a base as a substance that tends to accept them. By definition, for every acid, there

---

*Recall that pH is measured potentiometrically.

is also a *conjugate base*. That is, after an acid donates it proton to another base, the residual deprotonated moiety is called the acid's conjugate base. Thus, in the reaction

$$HA = H^+ + A^-$$

A^- is the conjugate base of the acid HA and is able to accept protons either from its parent or another acid.

Conversely, for every base, B, there is a conjugate acid. That is,

$$B + H^+ = BH^+$$

where BH^+ is the conjugate acid of the base B. Thus, ammonia, which readily accepts a proton, is a base in the reaction

$$NH_3 + H_+ = NH_4^+$$

with the ammonium ion being its conjugate acid.

But properties of acids and bases may only be manifested through their reactions with their counter species. That is, the definition of an acid makes sense only in the context of its interaction with a base. Thus, HCl (an acid) reacts with water (a base) in the *hydrolysis* reaction

$$HCl + H_2O = H_3O^+ + Cl^-$$

With Cl^- being the conjugate base of the hydrogen chloride and the hydrolized hydrogen ion, H_3O^+, being the conjugate acid of the water.

Acid and Base Strength

When an acid has a high propensity for giving up its proton, it is regarded as a strong acid, and its conjugate base is, therefore, weak. Thus, Cl^- in the hydrolysis reaction above is a weak base. Contrariwise, when a base strongly attracts protons, it is regarded as a strong base, and its conjugate acid is weak.

The measure of acid or base strength is manifested by the equilbrium established between the reactants and products of the hydrolysis reactions. That is, the further to the right the reaction

$$HA + H_2O = H_3O^+ + A^- \tag{10.1}$$

proceeds, the stronger is the acidity of HA relative to water.

Equilibrium reactions are characterized by an equilibrium constant, defined for eq. 10.1 as

$$K_{eq} = \frac{[H_3O^+][A^+]}{[HA][H_2O]} = \frac{[H^+][A^-]}{[HA]} \tag{10.2}$$

where the concentration (activity) of water has been taken as unity* and, for convenience, the symbol H_3O^+ has been replaced by its non-hydrated counterpart, H^+. Recalling the definition of $pH = -\log[H^+]$ and defining $-\log(K_{eq}) = pK_{eq}$, one can, therefore, with some re-arrangement, write

$$pH = pK_{eq} + \log\frac{[A^-]}{[HA]} \tag{10.3}\dagger$$

That is, the pH of an acid or base solution is dependent upon the ratio of the equilibrium concentrations of the acids and bases involved in the hydrolysis reactions. And, as can be deduced from eq. 10.3 and the definition of pKeq, the stronger the acid, the lower is its pK_{eq} value.

In aqueous systems, the strength of an acid or base is always referred to water. Consequently, water may be regarded either as an acid or a base, depending upon the acid strength of the substance with which it reacts. A substance more acid than water will contribute a proton to it forming the hydrolized hydrogen or *hydronium* ion, H_3O^+, and a substance more basic than water will will take a proton away from it forming the hydroxide ion, OH^-.

Ammonium ion, which, like hydrogen chloride, is a stonger acid than water, will, therefore, react with water according to

$$NH_4^+ + H_2O = H_3O^+ + NH_3$$

and acetate ion, which is a stronger base than water, will take protons from it to form hydroxide ions

$$CH_3COO^- + HOH = CH_3COOH + OH^- \tag{10.4}$$

In the case of very strong acids like HCl or very strong bases like NaOH, the equilibrium is, for all practical purposes all the way to the right. That

---

*In the definition of an equilibrium constant, as in the definitions of cell potentials, the concentration or, alternatively, the activity of a pure substance is taken as unity.
†Equation 10.3 is sometimes called the Henderson–Hasselbalch equation.

is, in the reactions

$$HCl + H_2O = H_3O^+ + Cl^-$$

and

$$NaOH + H_2O = Na^+ + OH^- + H_2O$$

only the ionic species exist in the aqueous environment, HCl and NaOH losing their identity as such.

Neutralization and Equivalence

As can be seen, the amounts of hydronium or hydroxide ions formed in the hydrolysis reactions are equivalent to the amounts of acidic or basic entities (analytes) that are undergoing hydrolysis. Because of equilibrium, if those ions are removed from the system, by neutralization by a counter species, for example, hydrolysis will continue until all of the hydrolizing base (or acid) is used up. Thus, in acidimetry and alkalimetry, the quantity of the appropriate counter species required to neutralize all the hydronium or hydroxide ions that can be formed from hydrolysis is also equivalent to the amount of base (or acid) in the original analysate. Consequently, for determination of acetate in an analysate, for example, one could titrate the hydroxide ions produced from the acetate's hydrolysis with an acid like HCl, continuing the titration and neutralizing all the hydroxide that is produced, and then use the amount of HCl titrated to determine the original acetate concentration.

The Ion Product of Water

Implied by the statement that the strengths of acids and bases in aqueous systems are always referred to water is water's ability to serve as both a proton donor and acceptor, i.e, behave as both an acid and a base, depending upon the material with which it interacts. Its ability to behave in both capacities leads to its interesting property of being able to serve as both a proton donor and acceptor *with itself*. That is, one water molecule may contribute a proton to another in the reaction

$$H_2O + H_2O = H_3O^+ + OH^-$$

This property* gives rise to water's ionic character, which is described

---

*Called autoprotolysis.

quantitatively by the equilibrium expression

$$K_{eq} = \frac{[H_3O^+][OH^-]}{[H_2O]^2}$$

Since the activity of water, as a pure substance, is defined to be equal to unity, this equation reduces to the well known relationship for the ion product of water

$$K_{eq} = [H^+][OH^-] = 10^{-14}$$

a relationship that holds in all aqueous systems, regardless of the nature of any other components that may be present.

Titration Curves

As mentioned earlier, the equivalence points of acid-base titrations may be determined by means of indicators that change color upon the addition of an amount of titrant slightly in excess of equivalence. Before the indicator changes color, however, and throughout the course of the titration, the pH of the system varies as a function of added titrant. The behavior of the resulting *titration curve* dictates first, whether a feasible titration may even be conducted and second, the type of indicator that must be selected to achieve a successful titration when one is possible.

As we shall see, the pH changes that occur when titrating strong acids against strong bases, strong acids against weak bases, and strong bases against weak acids are generally sharp at equivalence when the solution concentrations are sufficiently high overall. On the other hand, when weak bases and weak acids are titrated against one another, the pH changes that occur near their equivalence points are generally quite gradual [Kolthoff et al. 1969, 702]. Thus, weak acids and bases may not usually be titrated against one another with good precision.

Strong Acid/Strong Base Titrations

It is a straightforward exercise to calculate titration curves.* For example, if we assume the titration of a solution of 25 cc of 0.1 m sodium hydroxide, a strong base, with a solution of 0.1 m hydrochloric acid, a strong acid,

---

*In this example, we titrate an acid into a base. The arguments presented are identical if one is titrating a base into an acid, but the resulting titration curves would appear as mirror images.

the reaction of which is given by

$$HCl + NaOH = Na^+ + Cl^- + H_2O$$

we can calculate the pH for three different regions of the titration curve. The first is the region in which added HCl removes free OH^- from the sodium hydroxide solution and forms water. The pH in the first region may be calculated by inserting the concentration of the OH^- in the system into the expression for the ion product of water. So, for example, at the start of the titration, before any HCl has been added, the pH of the NaOH solution would be given by

$$[0.1][H^+] = 10^{-14}$$

$$pH = 13$$

At any point during the titration, the concentration of OH^- will be its residual concentration, i.e., the amount remaining in the solution divided by the original volume of the solution plus that of the added HCl. For our example system at a point, say, when 20 cc of HCl have been added, the pH is calculated as follows

$$OH^- = \text{Original amount of NaOH} = (0.1 \text{ m})(0.025 \text{ liters})$$

$$= 0.0025 \text{ moles}$$

$$OH^- \text{ remaining} = 0.0025 \text{ moles} - (0.02 \text{ liters})(0.1 \text{ moles/liter})$$

$$= 0.0005 \text{ moles}$$

$$OH^- \text{ concentration} = 0.0005 \text{ moles}/0.045 \text{ liter} = 0.0111 \text{ molar}$$

$$pH = \log[H^+] = -\log[(10^{-14})/(0.0111)] = 14 + \log(0.0111) = 12.04$$

One can, therefore, develop a curve for the first region of a titration by successively calculating OH^- concentrations as a function of HCl additions.

Near equivalence, however, as the concentration of sodium hydroxide sourced OH^- approaches zero, the ionization of the water produced by the titration will begin to determine the overall concentration of OH^- and H^+ in the system. And at equivalence, when 25 cc of the HCl has been titrated, the pH will be determined uniquely by the ionization of water, since all the hydroxide ions of the starting NaOH solution will have been converted by reaction with the added acid. At equivalence then, the concentration of H^+ will also be equal to the concentration of OH^-. In our example then, the pH at equivalence is given by

$$[H^+][OH] = [H^+]^2 = 10^{-14}$$

$$pH = -\log[H^+] = \log(10^{-7}) = 7$$

After equivalence, since no source of OH^- remains in the system, other than the slight amount produced by the ionization of water, the pH is determined entirely by the amount of excess HCl that is added.* So, for example, at a point at which 26 cc (1 cc in excess) of HCl have been added, the pH would be given by

amount of excess HCl = (0.1 moles/liter)(0.001 liters) = 0.0001 moles

concentration of H^+ = 0.0001 moles/0.051 liters = 0.00196 molar

$$pH = \log(0.00196) = 2.71$$

Completely developed titration curves for our example solutions, as well as for more dilute solutions of the acids and bases, are shown in Figure 10.1. As can be seen, though there is a sharp change in pH at the equivalence point for all concentrations, the net pH change at equivalence decreases as solution concentrations decrease. As will be seen in the discussion of acid-base indicators below, these differences in net pH change at equivalence, as well as the pH at equivalence, dictate the kind of indicators that may be selected for the titration.

Strong Acid/Weak Base Titrations and Vice Versa

Titration curves for strong acid/weak base titrations or strong base/weak acid titrations may be determined similarly. For example, if one was titrating a 0.1 molar solution of HCl against 25 cc of a 0.1 molar sodium acetate solution, the concentration of acetate ion would be reduced by the addition of the HCl. That is, referring to eq. 10.4, as HCl was added to the system, it would react with the OH^- ions produced by the hydrolysis of the acetate and form water. As it did, more acetate would hydrolyze to form more OH^- and more unionized acetic acid. For each mole of OH^- neutralized by the HCl, another mole of acetate would hydrolyze to maintain its base/conjugate acid equilibrium concentrations. Thus, there would be a one-for-one reduction of acetate ion for each mole of HCl added to the system. At equivalence, 25 cc of HCl would have converted all the acetate ion to acetic acid in the reaction

---

*The contribution of H^+ due to the ionization of water is negligible compared to the amount added and may, therefore, be ignored.

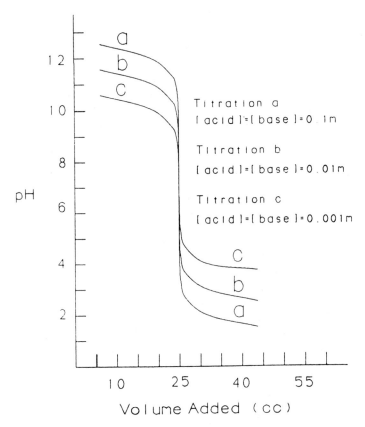

Figure 10.1   Strong acid:strong base titration curves. The concentrations indicated are those of both the acid titrant and the base being titrated. Note that the
starting pH values of the bases are higer for higher concentrations and the that
the ending pH values of the titrated systems are lowest for the most concentrated
acid titrants.

$$CH_3COONa + HCl = CH_3COOH + NaCl$$

forming 50 cc of acetic acid solution of concentration 0.05 molar.

But at equivalence, the only source of H + is from the hydrolysis of
the acetic acid produced. So, the H^+ concentration is determined by the
hydrolysis equilibrium constant (eq. 10.2)

$$K_{eq} = 1.74 \times 10^{-5} = \frac{[H^+][CH_3COO^-]}{[CH_3COOH]}$$

Also, since the only source of hydrogen ions at equivalence is the ionization of the acetic acid, the concentrations of H^+ and CH_3COO^- must be equal, and the equilibrium concentration of the acetic acid itself must be equal to the difference between the amount formed in the titration and the amount of it that ionized. So, one can write for the hydrolysis reaction

$$\frac{[H^+]^2}{[CH_3COOH] - [H^+]} = 1.74 \times 10^{-5}$$

Substituting the concentration of the acetic acid produced by the titration (0.05 molar in 50 cc) for its nonionized value, and solving the binomial expression for $[H^+]$, one gets 0.000924 for the value of the hydrogen ion concentration and, thus,

$$pH = -\log(0.000924) = 3.03$$

That is, the equivalence point for the titration of a strong acid against a weak base occurs well into the acid region, in contrast to that of a strong acid/strong base titration, where equivalence occurs at neutral pH values.*

To calculate the pH changes in the first region of the titration curve, in which the acetate ion concentration is being decreased by the added acid, one applies eq. 10.3

$$pH = 4.76 + \log \frac{[CH_3COO^-]}{[CH_3COOH]} \tag{10.5}$$

where 4.76 is the value of pK_{eq} for acetic acid. And, as can be deduced by examining eq. 10.5, as the acetate ion concentration is reduced (and the acetic acid concentration increases), the logarithmic term also decreases, resulting in a decrease in pH.

At the point at which the acetate ion and the acetic acid concentrations are equal, the pH becomes equal to the pK_{eq} for the hydrolysis reaction, and further titration reduces the pH to values approaching that at the equivalence point.

Beyond equivalence, the hydrogen ions introduced as excess HCl add to that resulting from the hydrolysis of the acetic acid produced before equivalence. The number of moles of H^+ in the system at equivalence is

---

*If we repeated the calculation for a strong base/weak acid titration, we would find that the equivalence point would fall well into the basic pH region.

given by

$$[H^+] \times Volume = \frac{0.000924 \text{ moles}}{\text{liter}} 0.05 \text{ liters} = 0.0000462 \text{ moles}$$

and each cc of excess HCl adds 0.0001 moles more. Thus for a 1 cc excess, the pH is given by

$$pH = -\log \frac{0.0001462 \text{ moles}}{0.051 \text{ liters}} = 2.54$$

and for a two cc addition, it is given by

$$pH = -\log \frac{0.0002462}{0.052} = 2.32$$

and so forth.

Typical titration curves for several weak bases are shown in Figure 10.2. The most significant feature of the illustration is the effect associated

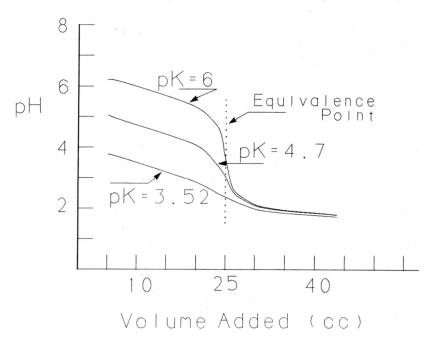

Figure 10.2 Strong acid:weak base titration curves. The curves are for a strong acid titrated against weak bases of varying base strengths. The pK values are for the conjugate acids of the bases.

with titrating bases of progressively increasing basicity, i.e., having conjugate acids of progressively decreasing acid strength and progressively increasing pK_{eq} values. As might be expected, as the base strength increases, the appearance of the titration curves approaches that of strong base–strong acid curves. At lower base strengths, however, the pH changes that occur at equivalence decrease and the curves flatten, reducing the resolution of the equivalence points as a function of titrant volume. For more acid substances, also, the value of the pH at the beginning of the transition is lower. Additionally, as with strong acid-strong base titrations, solutions of lower concentration lead to decreasing net pH changes at equivalence, as is illustrated in Figure 10.3.

Acid-Base Indicators

Acid-base indicators are generally weak organic acids or bases that change color over specific pH ranges. The basis of the color change resides in the reactions the indicators undergo. For example, a weak organic acid indicator, HIn, will be in equilibrium with its conjugate base, In, at any

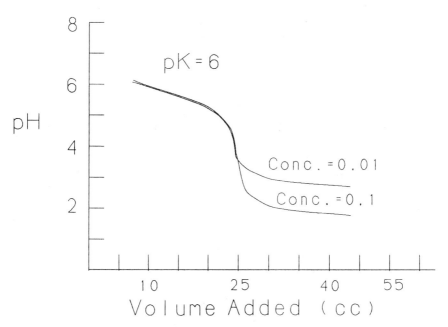

Figure 10.3   Concentration effects in titrating a strong acid against a weak base. The concentrations shown are for both the acid titrant and the base being titrated.

pH according to

$$HIn = H^+ + In^-$$

The pure acid form of the indicator will be one color while the pure basic form will be another. The equilibrium constant for the reaction is given by

$$K_{HIn} = \frac{[H^+][In^-]}{[HIn]}$$

and, rearranging, one gets

$$\frac{[In^-]}{[HIn]} = \frac{K_{HIn}}{[H^+]}$$

That is, the ratio of base to acid form and, thus, the equilibrium mix of color is dependent upon the pH of the system in which the indicator is placed as well as upon its hydrolysis equilibrium constant. Consequently, there is a finite pH range over which the color transition of the indicator occurs. *It is these factors that must be considered in selecting an indicator for a particular acid-base titration, the equivalence point of which may be either in the acid or base region and may extend over broad or narrow pH ranges, depending upon the acid-base strength relationships of the titrants and substances being titrated, as was discussed above.*

Detection of the Color Transition

It is interesting to explore the potential error of a typical acid-base titration by investigating what volume of acid, for example, would cause a visible color change to occur during a titration. Detection of a visible color change, of course, depends upon the viewer's color perception in the region of the spectrum that is being viewed. However, it has been determined that, in general, the physiological threshhold sensitivity to color is such that a distinction between a base color and one that is changed by the addition of another occurs when about 10% of the second is mixed with the first [Kolthoff et al. 1969, 691].

Thus, an observer would see the color of the pure acid indicator when

$$\frac{[In^-]}{[HIn]} = \frac{[K_{HIn}]}{[H^+]} < 0.1$$

and the color of the pure basic indicator when

$$\frac{[\text{In}^-]}{[\text{HIn}]} = \frac{[\text{K}_{\text{HIn}}]}{[\text{H}^+]} > 10$$

That is, the acid color would be seen when

$$[\text{H}^+] > 10[\text{K}_{\text{HIn}}]$$

and the base color would be seen when

$$[\text{H}^+] < 0.1[\text{K}_{\text{HIn}}]$$

Between these two values, an observer would see a gradation of color extending from one extreme to the other.

Expressed in terms of pH and pK_{HIn}, where pK_{HIn} is equal log K_{HIn}, we find that the approximate pH range over which a color change could be observed would be given by

$$\text{pH} = pK_{\text{HIn}} \pm 1$$

A typical acid-base indicator might have a pK_{HIn} value of about 5 (Table 10.1). Thus, a complete conversion from base to acid color requires enough excess acid to be added to change the analysate solution from a pH value of 6 to 4. That is, the hydrogen ion concentration of the analysate must be increased from 10^{-6} molar to 10^{-4} molar. This would require, for a typical acid titrant with a concentration of 0.1 molar being added to an analysate solution with a volume of 50 cc, an addition of about 0.5 cc of excess titrant. Since one usually tries to arrange titrations

Table 10.1   Approximate pK_{HIn} Values of Some Acid-Base Indicators and their Color Changes[a]

Indicator	pK_{HIn}	Color change
Thymol blue	2	red to yellow
Congo red	4	blue to red
Methyl orange	4	red to yellow
Ethyl red	5	colorless to red
Bromothymol blue	6.5	yellow to blue
Phenol red	7.5	yellow to red
Phenolphthalein	8.5	colorless to red
Alizarin yellow	11	yellow to red

[a]From Eastman Kodak Publication No. JJ-13, January, 1982.

so that about 25 cc of titrant are required to reach equivalence, the total color change would occur over an addition of about 2.5% of the total titration volume. But since the average observer can detect a color change just above the 10% color mix level, the typical acid-base titration may be expected to be reproducible, in so far as color detection is concerned, to about 0.05 cc or about the readability of a typical burette.

Oxidation-Reduction Titrations

Recall that oxidation-reduction reactions involve the transfer of electrons from a reducing agent (the substance being oxidized) to an oxidizing agent (the substance being reduced) and that each of the parties to the reaction have an associated half cell or Nernst potential given by

$$E = E_0 + \frac{RT}{nF} \log \frac{[A^{n+}]}{[A]} \tag{10.6}$$

where $[A]$ and $[A^{n+}]$ represent the concentrations of the reduced and oxidized states, respectively, of the redox agent. Recall also that the spontaneity of a redox agent discharging electrons to or accepting them from the external circuit of an electrochemical cell depends upon there being a difference in the Nernst potentials between the half cells.

In electrochemical cells, the components of the two half cells are generally separated by a porous membrane, allowing electron transfer to occur only through the external circuit. However, when two couples with different Nernst potentials are allowed to come in contact with one another, electron transfer also occurs. That is, a redox reaction ensues. Thus, one may titrate a reducing (or oxidizing) agent against a reducible (or oxidizable) analyte, and the equivalence point is the point at which essentially all the analyte is converted to its reduced (or oxidized) form.

The iodine/sulfur dioxide system described earlier in the section on Kinetic Methods in Ch. 8 is typical of systems that lend themselves to volumetric analysis by redox titration. Sulfur dioxide is readily oxidized to sulfate ions by elemental iodine, which is converted to iodide ion according to

$$SO_2 + I_2 + 2H_2O = SO_4^{2-} + 2I^- + 4H^+$$

In a typical analysis, a known quantity of iodine that is greater than that required to completely convert the sulfur dioxide is added to the system. Then the amount used for the conversion is determined by titrating the excess with sodium thiosulfate, an agent that reduces the excess iodine to iodide according to

$$I_2 + 2S_2O_3^{2-} = 2I^- + S_4O_6^{2-}$$

The equivalence point of the thiousulfate titration may be determined colorimetrically, as is discussed below, or one may use an iodide reversible electrode to monitor the change in iodide ion concentration in the system as the titration proceeds.

Oxidation-Reduction Titration Curves

Similar to acid-base titrations, one can construct theoretical redox titration curves by calculating the Nernst potentials for one or the other of the participants in the titration as a function of titrant addition. And similar to acid-base titrations, there are three stages of calculation, the first being before equivalence, the second at equivalence, and the third post equivalence. So consider, for example, a redox titration in which one titrates a 0.1 molar solution of Red_2 against 50 cc of a 0.01 molar solution of Ox_1 in the reaction

$$Ox_1 + Red_2 = Red_1 + Ox_2 \tag{10.7}$$

which involves the transfer of only one electron. We note that there are a total of 5×10^{-4} moles of Ox_1 in the solution being titrated.

The Nernst potential for the Ox_1/Red_1 couple is given by

$$E_1 = E_{01} + \frac{RT}{F} \log \frac{[Ox_1]}{[Red_1]}$$

Upon the addition of one cc of titrant, which in our example would contain 10^{-4} moles of Red_2, 10^{-4} moles of Ox_1 would be converted to Red_1. The potential of the Ox_1/Red_1 couple would then be

$$E = E_{01} + \frac{RT}{F} \log \frac{4 \times 10^{-4} \text{ moles}/0.051 \text{ liters}}{1 \times 10^{-4} \text{ moles}/0.051 \text{ liters}}$$

Clearly then, knowing the value for the standard potential for the system, one can progressively calculate other values of E for additions of titrant approaching the equivalence point.

The potential at the equivalence point is determined by applying the following argument. In the course of a titration, there is a one-for-one correspondence between the number of equivalents of titrant added to the system and the number of equivalents of the titrated substance that are converted to product. At the equivalence point, although there is essentially complete conversion of the titrated substance to its reaction product, a small concentration of both titrant and unreacted titrated material remain in the system, the amounts remaining being dependent upon the equilibrium relationship of eq. 10.7. But because of the one-for-one equivalence relationship of titrant and product, there is also a one for one relationship between residual titrant and unreacted titrated material. Consequently, $[Ox_1] = [Red_2]$ and $[Red_1] = [Ox_2]$, and one can write

$$\frac{[Ox_1]}{[Red_1]} = \frac{[Red_2]}{[Ox_2]}$$

The Nernst potentials for each of the couples at equivalence, therefore, become

$$E_{eq} = E_{01} + \frac{RT}{F} \log \frac{[Ox_1]}{[Red_1]}$$

and

$$E_{eq} = E_{02} + \frac{RT}{F} \log \frac{[Ox_2]}{[Red_2]} = E_{02} - \frac{RT}{F} \log \frac{[Ox_1]}{[Red_1]}$$

Adding the two expressions then yields

$$2E_{eq} = E_{01} + E_{02}$$

or

$$E_{eq} = \frac{E_{01} + E_{02}}{2}$$

Thus, the potential at equivalence is simply the average of the two standard potentials for the reacting redox couples.

After the equivalence point is reached, the concentration of Ox_2 remains essentially constant since there is no more Ox_1 to be reduced, and the potential of the system is dominated by additions of excess Red_2. Values post equivalence may, therefore, be determined from the expression for the Nernst potential for the Ox_2/Red_2 couple.

Typical redox titration curves are illustrated in Figure 10.4. The two curves represent potential changes in titrations involving couples with

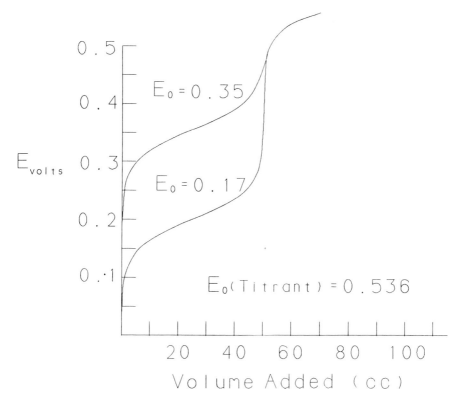

Figure 10.4   An oxidation-reduction titration showing the effects of standard potential differences on the behavior of the curve (see text).

standard potentials near to and somewhat distant from the standard potential of the titrant. As can be seen, the magnitude of the change in potential at equivalence is larger when the difference in the standard potentials between the titrant and substance being titrated is large. Naturally, a steeper change in potential at equivalence allows for a more precise determination of the equivalence point. Thus, to achieve reasonable analytical precision, it is generally recommended that titrants be selected for analysis of any particular system such that the difference between the standard potentials of the reacting components is greater than 0.2 volts [Peters et al. 1976, 207]. As will be seen below, the difference in standard potentials also dictates the feasibility of using a redox indicator to establish the equivalence point.

Oxidation-Reduction Indicators

In some oxidation-reduction titrations, the substance being titrated may serve as an indicator. For example, the iodometric titration discussed above may be carried out visually by adding a small amount of soluble starch to the titration medium. Starch forms an intense blue complex with iodine. Thus, the EPA prescribed method for the determination of sulfite in water [ASTM Annual Book, Method D 1339] is based on adding a known amount of iodine to an unknown along with a few cc of a 1% starch solution. The equivalence point is signalled by the disappearance of the blue color associated with the complex as the excess iodine is titrated with sodium thiosulfate solution.

But most titrimetric oxidation-reduction methods make use of oxidation-reduction indicators, a class of compounds whose oxidized and reduced forms differ in color. The behavior of a redox indicator may be represented by the Nernst equation

$$E = E_o + \frac{0.059}{n} \log \left[\frac{\text{Oxidized Form}}{\text{Reduced Form}} \right]$$

where the prelogarithmic term RT/F has been replaced by its numerical value at 25° C.

As with the acid-base indicators, a redox indicator will be the color of the oxidized form if its concentration is about 10 times that of the reduced form, and vica versa. That is, the potential at which the indicator's color will be that of the oxidized form will be given by

$$E_{ox} = E_0 + \frac{0.059}{n}$$

and the potential for that of the reduced form will be

$$E_{re} = E_o - \frac{0.059}{n}$$

Thus, the range over which the complete color change occurs is given by the difference between the two potentials or about 0.12 volts for a one-electron change.

Selection of a redox indicator, of course, is based on the criterion that its redox potential lies somewhere within the steep portion of the titration curve. Additionally, however, referring to the foregoing discussion, consideration must also be given to the magnitude of the difference in redox potentials of the titration system, selecting a titrant so that the difference is sufficiently larger than the color change range of the indicator. A general rule of redox titrations, therefore, is that the difference in redox potentials between titrant and analyte be of the order of 0.4 v, or about 3–4 times the color change range of a typical indicator [Peters et al. 1976, 207].

Advantages of Volumetric Analysis

Like gravimetric methods, volumetric methods, especially those in which indicators rather than instruments are used to detect equivalence points, require minimum investment in equipment. Additionally, relatively unsophisticated personnel can be rapidly trained to carry them out. Like gravimetric methods, though, volumetric analyses can be both time and labor intensive. However, there are automatic titrators available commercially that can reduce the labor factor in cases where analysis frequency is high. The action of an automatic titrator may be based on signals generated by pH, potentiometric, or amperometric electrode systems, or it may make use of the elements of a simple colorimeter to monitor an indicator color change. For the last, the titrated system in placed between a light source and an appropriately filtered photodetector. Feed-back from the detector system used controls the rate and amount of titrant added, and the titrant volume that is added to the equivalence point is thereby electronically recorded.

Environmental Applications of Titrimetry

Titrimetric methods have extensive applications in environmental analysis. Thus, methods for the determination of certain derivatives all of our subject elements have been prescribed by the EPA and are, as well, standard methods as defined by ASTM [ASTM Annual Book 1987] and the American Public Health Associations [Standard Methods 1981].

Titrimetric Methods for Carbon

Probably the most common application of titrimetry in carbon analysis is for determination of the acidity or alkalinity of waste and drinking waters. The former is caused by dissolved carbon dioxide, which forms carbonic acid, and the latter is caused by bicarbonate, carbonate, and hydroxide compounds. Carbon dioxide may be determined by titration with sodium hydroxide and, since both bicarbonate and carbonate ions are weak bases, they, along with hydroxide, may be measured by titration with a strong acid.

In a typical measurement of acidity [ASTM Annual Book, Method D-513E], the titration is carried to an equivalence point indicated by a phenolphthalein acid-base indicator, the color change of which occurs at about pH = 8.3. At that level of basicity, all the CO_2 has been converted to bicarbonate by the added base.

In a typical measurement of alkalinity [Golterman et al. 1978, 58], the titrant is hydrochloric acid. In this case two indicators may be used. The titration is carried first to a phenolphthalein color change, at which point all the carbonate ions in the sample have been converted to bicarbonate and all the OH^- has been neutralized. Thus, the quantity of CO_3^{2-} and OH^- ions may be separated from other alkalinity-contributing components by means of the titrant volume used to get to the first color change. Continued titration thereafter to an equivalence point at pH 4.3, indicated by a methyl orange indicator, permits determination of the total amount of alkalinity, because pH 4.3 is the point at which all the HCO_3^- has been converted to carbon dioxide.

The EPA-prescribed acidity and alkalinity methods call for titrations directly to a phenolphthalein equivalence point with NaOH for measuring acidity, and titration directly to a methyl orange equivalence with HCl for measuring alkalinity [40 CFR 1987, Part 136.3].

Chemical oxygen demand, which we have mentioned in earlier sections of this monograph, is determined via a redox titration in which the strong oxidizing agent potassium dichromate is used to oxidize organic carbon in water to carbon dioxide and water. Similar to iodometric titrations, a quantity of K_2CrO_4 larger than the expected amount of oxidizable substance in the sample is used, and the excess is titrated with a reducing agent to determine the amount used up. In one EPA approved method [ASTM Annual Book, Method D 1252], the excess dichromate is titrated with a solution of ferrous ammonium sulfate with 1,10-(ortho)-phenanthroline monohydrate being used as the redox indicator.

Titrimetric Methods for Nitrogen

As has been discussed before, nitrogen pollutants in water may exist in a number of forms including nitrites, nitrates, organic nitrogen, and ammonia, with the last usually present as ammonium salt. The last two of these lend themselves to analysis by acid-base titrimetry, and a number of titrimetric methods have received EPA approval.

Organic nitrogen may be determined by means of a very important and commonly used titrimetric procedure known as the Kjeldahl method. Kjeldahl nitrogen determinations are based on the catalytic conversion*

---

*The sample is boiled in concentrated sulfuric acid to which either mercuric sulfate or a mixture of copper sulfate and selenium metal is added to serve as the catalyst.

of organic nitrogen to ammonium sulfate, followed by conversion of the ammonium sulfate to ammonium hydroxide by reaction with sodium hydroxide. Heating highly basic ammonium hydroxide solutions can free ammonia according to

$$NH_4OH = NH_3 + H_2O$$

So, after treatment with NaOH, the resulting solution is distilled to free the ammonia, which is collected in a solution of boric acid and titrated with hydrochloric acid. The details of the Kjeldahl method may be found in virtually any textbook on analytical chemistry; the EPA approved method is described in ASTM Method D 3590 [ASTM Annual Book 1987].

An EPA approved method to determine non-organic, ammonia-associated nitrogen in water is described in ASTM method D 1426 [ASTM Annual Book 1987]. In this procedure, the sample is made alkaline to convert the ammonium salts to ammonium hydroxide and, similar to the Kjeldahl protocol, the ammonia is then distilled into a boric acid solution where it is titrated with a strong acid.

Reported sensitivities of titrimetric determinations of nitrogen are of the order of $0.1 - 1$ mg/l.

Titrimetric Methods for Sulfur

Determination of sulfur dioxide and sulfite salts via iodometric titration has been discussed a number of times. The EPA approved method [ASTM Annual Book, Method D 1339] for determination of sulfite in water, however, is based on a modification of the reactions discussed earlier. The method is based on the generation of I_2 by an iodate salt in an acid solution containing iodide ions. That is, the reaction

$$IO_3^- + 5I^- + 6H^+ = 3I_2 + 3H_2O$$

provides the iodine used to oxidize the sulfite in the sample according to

$$SO_3^{2-} + I_2 + H_2O = SO_4^{2-} + 2I^-$$

In the procedure, an iodate solution is titrated into an acidified sample to which potassium iodide and soluble starch have been added. As long as sulfite is present, it converts the I_2 produced by the reaction of iodate and iodide to additional iodide ion. However, when the sulfite is consumed, iodate added to the sample solution produces excess I_2, which combines with the starch to produce the characteristic color of the starch-iodine complex. The amount of iodate titrated, therefore, is stoichiometrically equivalent to the sulfite in the sample.

Sulfides in water may also be determined iodometrically. In the EPA approved procedure [Standard Methods 1981, Method 427D], a sample is acidified to convert the sulfides to H_2S, which is then purged with an inert gas into an acidified solution of zinc acetate. Iodine solution is added to the zinc sulfide suspension thus produced, oxidizing the sulfide to sulfate. Titration of the excess iodine remaining in the system with sodium thiosulfate is then used to quantitate the sulfides in the original sample.

And, as a last example, a titrimetric method for determining the level of sulfur dioxide in working place atmospheres is described in ASTM Method D 3449. The method involves sampling the air space by drawing a fixed volume through a solution of hydrogen peroxide, oxidizing any sulfur dioxide present to sulfate. The sulfate thus produced is then titrated with a solution of barium perchlorate to an equivalence point determined with a thorin* indicator.

GAS DETECTION TUBES

A convenient, though not a highly precise way to monitor hazardous gases and vapors in the environment is by means of gas detection tubes. These devices, which are available commercially from such companies as Matheson Gas Products in the U.S. and Dragerwerk A.G. in West Germany, are thin glass tubes that are packed with reagents that react specifically with the gases being monitored to produce colored products. They are used commonly to monitor working-place environments as part of occupational health and safety programs.

The operational principles of gas detection tubes are a kind of blend of reflectance colorimetry and gas chromatography. The packings are usually made of an inert powdered substrate onto which the gas specific reactant is coated. In use, a fixed volume of air is pumped through the tube, with analyte vapor in the air sample reacting with the coating as it flows down the tube's length. As it does so, the color of the coating changes. But because the amount of color-forming reactant per unit length of tube is limited, excess analyte vapor is able to traverse longer distances through the length of the tube before being consumed. Consequently, the length of the colored section of the tube becomes proportional to the amount of analyte vapor in the air sample.

---

*The chemical name for thorin is o-[3,6-disulfo-2-hydroxy-1-napthylazo] benzene arsonic acid.

Table 10.2 Some Gases and Vapors for which Gas Detector Tubes are Available

Analyte	Sensitivity
Ammonia	0.05–1 vol.%
Benzene	0.5–104 ppm
Carbon dioxide	0.1–1.2 vol.%
Carbon disulfide	0.1–10 mg/L
Carbon monoxide	2–60 ppm
Diethyl ether	100–4000 ppm
Dimethyl sulfide	1–15 ppm
Ethylene oxide	1–15 ppm
Formaldehyde	0.2–2.5 ppm
Hydrazine	0.25–5 ppm
Hydrocarbons	0.1–1.3 vol.%
Hydrogen sulfide	0.5–15 ppm
Nitrogen dioxide	0.5–10 ppm
Pyridine	5 ppm
Sulfur dioxide	0.1–3 ppm
Vinyl chloride	0.5–3 ppm

As an illustration of how the amount of coating placed on tube's packing is determined, we will consider a tube designed to detect hydrogen sulfide in the range of one to $20\,\mu l$ per liter of air. The color reaction is based on the formation of black lead sulfide as a result of the reaction of lead acetate with hydrogen sulfide according to

$$Pb(C_2H_3O_2)_2 + H_2S = PbS + 2HC_2H_3O_2$$

At standard temperature and pressure, one mole or 22.4 l of H_2S will react with one mole or 207.2 g of lead acetate. Thus, in our example tube, which is designed to detect a maximum of $20\,\mu l$ of H_2S in a one liter air sample, one would require about $173\,\mu g$ of lead acetate. Such an amount is coated on about a gram of inert substrate and loaded into a tube 3–4 mm in diameter and about 10–12 cm in length. About half of such a tube would turn black upon passing a one liter sample of air containing about $10\,\mu l$ of H_2S through it.

Gas detection tubes are available for several hundred different toxic gases and vapors. A partial list of some of them along with typical sensitivities as claimed by one manufacturer* is shown in Table 10.2. As can be

*Dragerwerk AG, Lubeck, West Germany publishes a handbook of Detector Tubes which lists all those available as well as their sensitivities and methods of use [Leichnitz 1985].

seen, tubes are available for a large number of carbon-containing gases as well as for the oxides of sulfur and nitrogen.

The analytical precision and accuracy of detection tubes are limited by the reproducibility and uniformity with which the gas specific reactant may be coated on the substrate. Both factors will affect the length of the reactant that is affected by the incoming analyte vapor. Additionally, the tubes have limited shelf lives, usually no more than two years, the gas specific reactants degrading in that time period. Manufacturers' claims for tube reproducibility vary among detector tubes, varying between values of the order of 10–15% for tubes recommended for quantitative analysis to just simple indication of the presence of the gas being analyzed for tubes recommended for only qualitative analysis.

It is probably for reasons of precision and accuracy that gas detection tubes have not generally been approved for regulated monitoring functions by EPA or the U.S. National Institute of Occupational Safety and Health. They are, however, useful devices for rapid testing of working environments, as long as results obtained with them are regularly verified by means of approved methods. Recommended gas detection tube practices are outlined in ASTM Method D 4490 [ASTM Annual Book].

COMBUSTION METHODS

As in the original fashion of Lavoisier, combustion analysis involves burning a sample in oxygen, generating its product oxides, and relating their amounts to the composition of the analysate. All three of our subject elements may be thus determined in most solids and liquids. However, a primary application for combustion analysis in environmental monitoring is for determination of organic carbon in waste water.

Principles

Combustion analysis has its foundations in the determination of the elemental composition of organic compounds. Its formalism dates from the middle of the 19th century when Liebig and Dumas developed many of the methods used to this day to determine the carbon, hydrogen, and nitrogen compositions of organics. In the early part of this century, however, major modifications were made to the methods of Liebig and Dumas by Fritz Pregl as he developed combustion methods specifically designed for the analysis of very small samples [Roth 1937]. Pregl, who won a Nobel prize for his work, developed his methods in response to the needs of the synthetic organic chemists of his time, whose research activities then could produce only very small quantities of new synthetic compounds. Because the analytical methods available to them, in general, destroyed their new materials, these early synthesis workers were loathe to give up any more than absolutely necessary. Thus, Pregl's micro-methods were a welcome development then and remain a mainstay of modern combustion analysis as well.*

The principle underlying every combustion method, macro or micro, is the same. That is, the sample is burned in an oxygen environment and the products are measured. A schematic of a typical combustion analyzer is shown in Figure 10.5. There is, however, a wealth of literature describing countless variations on the basic schematic, the variations involving the physical configuration of the apparatus, the temperature of the combustion

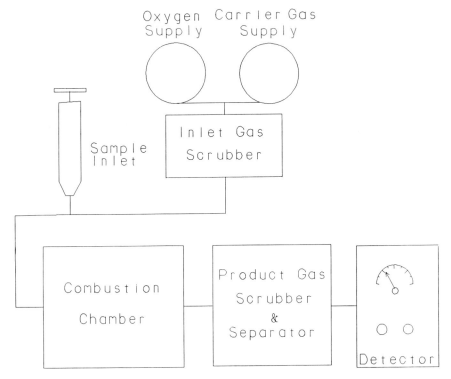

Figure 10.5 A general schematic of a combustion analyzer.

---

*Pregl's micromethods have the additional advantage of allowing combustion analysis of virtually any kind of sample, including highly flammable and even explosive materials. This is because the small samples used lend themselves to controlled burning.

chamber, the source of oxygen and the method used to combine it with the sample, and the manner in which the combustion products are separated, purified, and measured [Ma and Rittner 1979; Dixon 1968; Bance 1980].

The host of different approaches used in combustion analysis implies a great deal of art in addition to science that is applied to any particular method described in the literature. This is verified, for example, in the variations on Pregl's procedures that may be found for the determination of carbon and hydrogen. Thus, Pregl originally recommended vaporizing a sample and driving it with a stream of pure oxygen into a packing of copper oxide held at about 700° C. Copper oxide, incidentally, was Liebig's selected oxidant. However, other workers have reported the use of barium chromate or silver permanganate as alternatives. Modifications of Pregl procedures have also included burning the sample directly in an oxygen stream in an open combustion tube held at about 900° C as well as simply burning the sample in a closed flask containing enough oxgen to convert it into its oxides.

Sulfur analysis has also been carried out in open tubes and closed flasks in an atmosphere of oxygen, burning the analysate and converting the sulfur in it to SO_2. And nitrogen has been determined similarly, with most procedures, however, being based on Dumas' original method wherein the sample is intimately mixed with copper oxide, which serves as the oxygen source for combustion in a carbon dioxide environment. The nitrogen oxide end products of this combustion process are then swept with the carbon dioxide into a second combustion chamber, which contains a packing of metallic copper, where they are reduced to elemental nitrogen and then measured as such in some appropriate way.

Gas scrubbing procedures are essential components of all combustion methods, especially on the scale of microchemical analysis. For example, purification of the oxygen used for burning is a necessity, and scrubbing impurities that always accompany combustion processes out of the desired product gases is also frequently a necessary part of a particular protocol. Thus, the literature of combustion analysis is replete with scrubbing methods [Ma and Rittner 1979; Methods in Microanalysis 1978]. Some examples include purifying incoming oxygen by passing it through an ascarite* trap to remove carbon dioxide and through anhydrone† to remove water in systems designed to measure carbon and hydrogen. Removal of interfering gaseous halides and sulfur and nitrogen oxides from the product stream may be accomplished by passing it through secondary heated zones

---

*Sodium hydroxide coated on a high surface inert adsorbent like asbestos.
†Anhydrous magnesium perchlorate.

containing metallic copper and silver. The copper, of course, reduces the nitrogen oxides to elemental nitrogen, which may be vented without measurement; and the silver packing removes halide and sulfur oxide impurities by converting them to silver halides and sulfides.

In still other procedures, the oxides of nitrogen are removed by passing the product stream through a trap containing manganese dioxide, which selectively absorbs them; and in analyses for nitrogen or sulfur, water and carbon dioxide impurities may be removed in the same way that they are scrubbed from incoming oxygen, i.e., by passing the product stream through ascarite and anhydrone traps.

A host of different methods may also be used to quantitate the combustion products, many of which we have already discussed. Gravimetric and volumetric analyses are frequently applied. Thus, carbon dioxide may be determined gravimetrically by absorbing it on ascarite and weighing the product. Similarly, water may be determined by absorbing it on anhydrone. Elemental nitrogen generated in Dumas type determinations can be measured volumetrically, by displacement of mercury, for example. And sulfur dioxide may be measured by passing the product stream into a solution of hydrogen peroxide, where the SO_2 is oxidized to sulfuric acid; quantitation is then achieved by acid-base titration or by barium sulfate precipitation.

Most modern combustion analyzers, however, make use of more sophisticated instrumented methods. Thus, gas chromatography has been used to separate product gases, quantitating them by means of thermal conductivity or flame ionization detectors; and conductimetry, coulometry, potentiometry, NDIR, and mass spectrometry, among others, have been variously applied to determination of carbon and sulfur [Ma and Rittner 1979, 48].

Typically, commercial CHN (carbon-hydrogen-nitrogen) analyzers will use the gas absorption methods mentioned above to separate the product gases and then measure them individually by passing them through thermal conductivity detectors. Thus, in such a typical analyzer, the product gases, after impurities have been scrubbed, are swept by an inert carrier like helium through a series of three thermal conductivity cells separated by traps. The first thermal conductivity cell provides a measure of the total amount of product gas. The trap between the first and second conductivity cells removes the water. Consequently, the second conductivity cell measures the carbon dioxide and nitrogen in the product stream. Finally, a second trap is used to remove the carbon dioxide, and the third conductivity cell measures the nitrogen. Hence, by difference, quantitation of all three of the gases is achieved.

Coulometric titration is used in a number of different commercial combustion analyzers to quantitate both sulfur and carbon. For the former, the SO_2 produced from the combustion is iodometrically titrated using a coulometer to produce the iodine titrant. Coulometric carbon analysis is based on sweeping the carbon dioxide produced by the combustion into a cell that contains a solution that reacts with CO_2 to form a titratable acid. The acid is then titrated with a coulometrically generated base to an equivalence point detected photometrically (see section on Coulometry in Ch. 8).

Non-dispersive infrared detectors (NDIR) are also commonly used for quantitation of carbon dioxide, because NDIR is one of the EPA sanctioned detection methods for determination of total organic carbon via combustion analysis.

Environmental Applications of Combustion Analysis

By far and away, the most common application of combustion analysis in environmental monitoring of our subject elements is for the determination of organic carbon in waste water. In fact, organic carbon's determination in water by combustion* is rapidly replacing BOD and COD as the recommended method for estimation of oxidizables, the bulk of which in waste water are present in the form of organic carbon. Combustion methods are also used to analyze the elemental compositions of particulates collected on filters in air and water pollution investigations, and the sulfur content of coal and petroleum fuels is determined by combustion analysis as an aid in the control of sulfur emissions from power plants.

Carbon in Water

Despite the availability of very sophisticated GC-MS methods for the analysis of water, total carbon levels remain the generally accepted criterion for classifying water quality. This is because although over 300 organic compounds have been identified by means of the more sophisticated approaches, the family of identified compounds represents only 5–10% of the carbon pollutants that may be present in contaminated water.

Carbon exists in water in both inorganic and organic forms, the former being primarily as carbonate salts. Thus, a distinction is made in water analysis among the values of *total carbon* (TC), *total organic carbon*

---

*See below for a discussion of u.v. catalyzed oxidation as an alternative to high temperature combustion.

(TOC), and *total inorganic carbon* (TIC). In environmental applications, total organic carbon is often further divided into several other categories. These include *purgable organics* (POC) and *nonpurgable organics* (NPOC), the difference between the two being those that are volatile enough to be purged from the water by passing a gas through it and those that are not. Dissolved carbon (DC), which is the residual after particulate matter is filtered from the sample is still another category as is dissolved organic carbon (DOC), which is the residual after filtration and removal of the inorganic forms by acidification.

Commercial carbon analyzers usually make provision for determination of TC, TOC, and TIC. Although specific protocols depend upon the particular analyzer selected, in a typical procedure, a water sample will be first purged with pure oxygen; this frees the volatile organics, which are swept by the oxygen through the analyzer where they are burned and POC is measured. The sample will then be acidified to convert carbonates to carbon dioxide, which is routed to the detector of the analyzer to determine TIC. And, finally, a portion of the carbonate- and purgables-free sample will be injected directly into the hot zone of the analyzer, where the water flash evaporates and the nonpurgable organics are burned to CO_2 for determination of NPOC. In the most modern analyzers, the various data are automatically combined to provide values for TC and TOC in an output report.

As mentioned, NDIR is the measurement method of choice in most commercial carbon analyzers because of its approval by the EPA. However, EPA also recognizes a measurement approach in which the carbon dioxide produced from the combustion process is mixed with hydrogen gas and then is catalytically reduced to methane by passing it through a post-combustion chamber packed with nickel and held at a temperature of 350° C. The methane is then measured with a flame ionization detector. Both the NDIR and flame ionization methods are described in ASTM Method D 2579 [ASTM Annual Book], which is also an EPA prescribed high temperature combustion procedure for measurement of TOC in municipal and industrial waste waters [40 CFR Part 136.3].

The sensitivity of combustion methods for determining organic carbon in water is limited to the order of ten ppm. The limitation is not fundamental to the instrument or the method. Rather it derives from the size of the water sample that may be burned. That is, only a very small sample may be injected into a high temperature furnace. To do otherwise is to invite flash evaporation with explosive intensity. The usual sample size for TOC determinations, therefore, is of the order of a few hundred microliters which, at the low end of the sensitivity scale, corresponds to several micrograms of carbon.

Sulfur in Fuels

A number of analyzers devoted specifically to determination of sulfur in fuels are available commercially, the major distinctions among them being the methods used for measurement and the temperature at which the sample is burned. Combustion temperature is an important factor in sulfur analysis. It can dictate the manner in which the instrument is standardized as well as the manner in which one interprets the analysis results. That is, combustion temperature influences the chemical nature of the sulfur oxides produced. Combustion at temperatures above about 1200° C tends to favor production of all sulfur dioxide while burning at lower temperatures tends to produce mixtures of SO_2 and SO_3.

When a mixture of sulfur oxides is produced by an analyzer, a more complicated instrument calibration protocol is often necessary. This is because the ratio of SO_2 to SO_3 that is generated at lower combustion temperatures is frequently dependent upon the chemical and physical properties of the burned sample as well as upon the combustion temperature. That is, the SO_2/SO_3 ratio produced is said to be *matrix dependent*. Thus, in systems in which the product gases are mixtures of sulfur oxides, it is necessary to calibrate the instrument with a known that is similar to the unknowns being analyzed, a requirement that can complicate testing of samples from diverse sources.

In some analyzers operated at lower temperatures, attempts are made to assure complete conversion of the sulfur oxides produced to all SO_2. One way this is done is by passing the product gases through a pair of post-combustion zones maintained at about 1000° C. In the first of the pair, which contains copper oxide, the sulfur oxides are converted to all SO3, and in the second, metallic copper reduces the SO_3 to SO_2. Such post-treatment, thus, allows the instrument to be calibrated by simply passing standard samples of sulfur dioxide through the detector.

Detection of sulfur dioxide may be accomplished in a number of ways. Coulometric titration and SO_2 fluorescence, both of which have been discussed earlier, provide the advantage of being highly specific. However, a number of perfectly reliable commercial analyzers make use of NDIR detection as well as thermal and electrolytic conductivity. In the last, the sulfur dioxide generated is bubbled through a dilute solution of sulfuric acid and hydrogen peroxide, where the SO_2 is oxidized to form additional hydrogen and bisulfate ions, thus increasing the solution's conductivity [Symanski and Bruckenstein 1986]. Finally, in the very well known Fisher Sulfur Analyzer* which, incidentally, may be operated at

---

*Allied Fisher Scientific Co., Springfield, N.J.

very high temperatures ($1500°$ C), SO_2 is measured iodometrically with an automatic titrator, using an amperometric sensor that controls the release of titrant from an automatic burette.

Nitrogen Analysis

Recognition that nitrogen oxides also contribute to acid rain has stimulated increased interest in measuring the nitrogen content of fossil fuels. Elemental analyzers of the kind discussed above, which can measure all the oxides produced from sample combustion are certainly suitable for the task. However, there are also combustion analyzers that are specifically designed for the determination of nitrogen in solids by virtue of the kind of detector incorporated into them. Thus, one commercial analyzer[*] makes use of a chemiluminescent detector specific for nitrogen (see section on chemiluminescence) in tandem with standard combustion apparatus.

LOW TEMPERATURE CHEMICAL OXIDATION METHODS

Particularly applicable to determination of TOC, chemical oxidation methods may in some cases remove the sensitivity limitations imposed by sample size in high temperature oxidation methods. In chemical oxidation, organic carbon in water is oxidized to CO_2 by treating the sample with a solution of a persulfate salt.[†] Because the oxidation may be carried out at low temperatures, large samples may be analyzed, producing proportionately larger quantities of carbon dioxide and enhancing sensitivity.

In some older manual methods, the oxidation is carried out in a sealed ampule that is heated to a temperature above the boiling point of water (approx. $150\,°C$). At the completion of the oxidation process, usually after several hours, the ampule is opened inside of a gas flow system that sparges the carbon dioxide produced out of the sample and sweeps it through drying tubes and into a detector to quantitate the carbon content.

A number of commercial instruments, however, take advantage of a more efficient oxidation process by exposing the sample-persulfate mix to high intensity ultraviolet radiation. The radiation catalyzes the oxidation process, reducing the analysis time to a few minutes. Equipment for determination of TOC by uv-catalyzed oxidation usually makes use of apparatus similar to conventional combustion apparatus. Instead of a combustion tube, however, the sample is injected into a reaction vessel containing a solution of the persulfate salt. The vessel may be made of

---

[*]Rosemount Analytical Division, Dohrmann Corp., Santa Clara, CA.
[†]$Na_2S_2O_8$ for example.

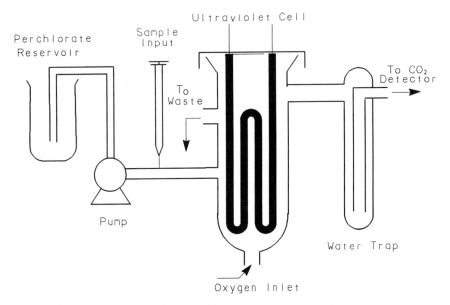

Figure 10.6 A schematic of an ultraviolet catalyzed chemical oxidation analyzer.

an ultraviolet transparent material like quartz, with the radiation source external to it, or the radiation source may be inserted directly within it (Figure 10.6). The oxidized sample is sparged from the reaction chamber with oxygen and passed through scrubbers to remove water. It is then routed through a detector for measurement of the CO_2 produced.

Despite the advantages of chemical oxidation with respect to potential sensitivity, however, it suffers some disadvantages relative to ordinary combustion methods. The oxidation times for the older manual methods are quite long, and ultraviolet catalyzed oxidations can not be used to measure TOC in samples containing high levels of particulates or halide salts; both are efficient uv attenuators. It has also been reported that chemical oxidation is not able to oxidize certain so-called refractory, i.e., difficult-to-oxidize organics, that are constituents of the non-purgables found in waste waters [Bernard 1985].

CONTINUOUS FLOW ANALYZERS

The recent advent of laboratory robots that are able to emulate the manipulations of human laboratory technicians has highlighted the cost-saving potential and enhanced reliability of automated chemical analysis [Laboratory Robotics Handbook 1988]. What is sometimes lost in the promotional activities associated with robots in the laboratory, however, is that automated chemical analysis has been around for over a generation, having been introduced originally in 1957.* These automated methods were manifested in the form of the continuous flow analyzers that were originally developed to accomodate the needs of clinical chemistry laboratories where, in typical mid-sized to large hospitals, monthly analytical requirements can involve multiple assays on thousands of specimens. However, the usefulness of continuous flow analysis rapidly spread to academic and industrial laboratories as well, that is, wherever large numbers of repetitive analytical determinations were required.

Principles

Continuous flow methods are characterized by apparatus that makes use of a robotic pipette to automatically aspirate analysate samples from discrete sample containers and inject them at a fixed frequency into a carrier stream. The carrier stream may contain the transformation reactants involved in the conversion of the analyte to a measurable product, or the stream may move the analysate through the analyzer with transformation reactants being merged with it at appropriate stations. The product of the conversion reaction that proceeds within the stream as the stream moves through the analyzer is then measured as it exits the system. Thus, the output from a continuous flow analyzer is a series of perturbations from a base line signal, each of the perturbations reflecting the passage of analyte through the detector, and with the magnitude of any perturbation representing a measure of an analyte's concentration in the originally injected analysate sample.

Modern continuous flow analyzers are of two types, those that make use of air-segmented carrier streams, with the analysate being carried in small elements of fluid interspersed between the air segments, and those in which the carrier stream is continuous, with the analysate occupying a discrete reaction "zone" in the stream. The first is represented by the well known "AutoAnalyzer" marketed by the Technicon Corporation (Tarrytown, N.Y.), and the second is represented by the family of instruments known generically as "Flow Injection Analyzers."

Air segmentation in AutoAnalyzers, which is accomplished by injecting small increments of air into the carrier stream at timed intervals, helps to maintain sample integrity by keeping a level of separation between

---

*The Technicon Corporation introduced its "AutoAnalyzer" in 1957.

samples, and it inhibits analyte carry-over between samples by having the air segments "wipe" the flow tube as they traverse the system. In the case of flow injection analyzers, sample integrity is maintained by fixing the geometry of the system and the flow rate such that samples are reacted and measured in a time period small compared to the time it might take for two adjacent samples to mix.

In air-segmented flow analyzers, the increments of reacting materials flowing between the air segments behave as small independent reaction chambers. The timing from sample input to detection and measurement is such that the analytical reactions that occur within these small chambers may proceed to completion.

In contrast, in flow injection analysis, the timing is such that reaction completion is rarely achieved before measurement [Ruzicka and Hansen 1981, 11]. However, it is an established fact of analytical chemical measurement that completion of a conversion reaction is not an absolute requirement for doing quantitative analysis, just so long as reaction conditions are identical for all the samples and calibration standards being processed. When that is the case, the detector responses are proportional to the analyte concentrations in each of the samples and knowns passing through the system, regardless of the extent to which the operating conditions permit the conversion reactions to proceed to completion. Thus, an analyte may be determined by comparison with the detector's response to that of a known.

A schematic showing the flow characteristics in both types of instruments is shown in Figure 10.7. As is illustrated, in the segmented system, the flow stream is "debubbled" before it enters the detector cell so as to eliminate noise signals that might be generated by the air segments. The resulting signal is, then, a series of pulses corresponding to the analyte concentrations in each of the reaction chambers that had been separated by the air segments; for any single sample, therefore, the total signal profile is a series of pulses of increasing and then decreasing magnitude. In contrast, the signal generated in a non-segmented system is more like that of a chromatograph, with the magnitude of a single pulse representing the concentration of the analyte.

In both types of analyzers, timing and reaction conditions are controlled by appropriate combinations of flow rate, the transit length through which the carrier stream flows, and treatment modules through which the stream passes in its transit to the measurement device. Delays in flow to allow for extended reaction times are incorporated by inserting reaction coils in the stream, the lengths of which are adjusted to achieve the desired timing. Heating and cooling, if necessary for specific protocols, are accomplished by controlling the temperature of the reaction coils; and rather sophisticated treatment modules have been developed that allow

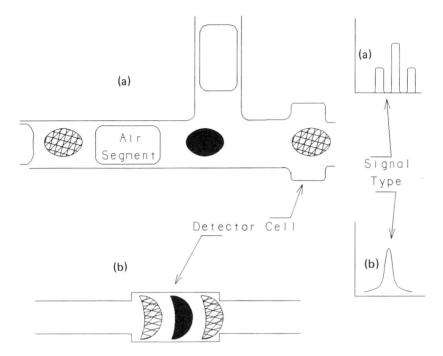

Figure 10.7   Flow characteristics in segmented and non-segmented continuous flow analyzers. The air bubbles in the segmented analyzer are removed from the stream before it enters the detector cell. The grids and filled symbols denote lower and higher concentration levels, respectively, in the stream.

for such processes as sample filtration, dialysis, liquid-liquid extraction, and gas diffusion to be carried out.

Dialysis modules consist of a pair of flow chambers separated by a dialysis membrane. As the sample stream passes through one side of the module, the analyte dialyzes through the membrane and is collected by the stream on the opposite side, thence being transported through the system for further processing and/or measurement.

Gas diffusion out of a sample stream, for the measurement of carbon dioxide, for example, may be accomplished similarly, in a module that allows the CO_2 to diffuse through a gas permeable membrane into a collector stream whose pH is being monitored.

Extraction is accomplished by combining the sample stream with a non-miscible extractant stream and passing the combination into an extraction coil where the two fluids are maintained in intimate contact as they flow through it. At the exit of the extraction coil, the fluids are passed

into a vessel where they can separate, and the one containing the analyte
is passed on for further processing or measurement.

Common to both segmented and nonsegmented continuous flow analyzers is the peristaltic pump used to drive the fluid streams through
them, a schematic of which is shown in Figure 10.8. As can be seen, the
pump is a positive displacement device based on the action of a set of
steel rollers that roll across a length of elastic tubing immobilized on a
flat surface called a platten. As they move, the rollers compress the
tubing and force the fluid to move in front of them with a resulting flow
rate that is dependent upon the tubing diameter and the roller velocity.

A major asset of the peristaltic pump in continuous flow analyzers is
that one pump may be used to control the flow rates of a number of
different fluid streams by simply paralleling tubes of different diameters
on the platten. Since the rollers move over all the tubes at the same
velocity, varying flow rates may be obtained because the flows in each of
the tubes are then dependent only upon their diameters. Thus, various
fluid rate ratios may be established in the analyzer. The same pump may
be used, for example, to control the flow rates of analytical reagents,
sample streams, collector streams when filtration or dialysis modules are
used, and to draw air for injection into segmented systems as well as to
withdraw it previous to the stream's entry into the detection device.

Where sequential reactant additions are necessary for a specific analytical protocol, they are merged into the carrier stream as it traverses the

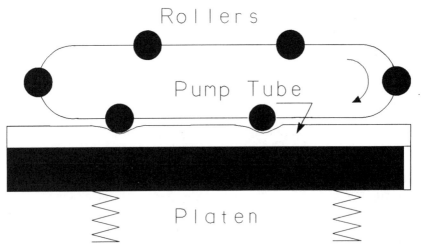

Figure 10.8   A peristaltic pump.

system. Timing between additions is controlled by inserting delay coils between the merge points. And, since the volume ratios of the reactants and flow stream components that are merging together are exactly equal to their flow ratios, volumetric additions of reactants may be controlled by paralleling appropriately sized reactant and flow stream tubing on the pump platten.

Virtually every measurement method used in manual analytical procedures has been adapted to continuous flow analyzers. Thus, colorimetry, potentiometry, amperometry, coulometry, and most spectroscopic techniques have been applied, the requirement for any particular detector being only that it be adaptable to small volume flow cells or that a method be available to extract sequential representative samples of the flowing stream for injection into it. The most commonly used detection method, however, is colorimetry.

To illustrate the processes underway in continuous flow analyzers, we use as an example the determination of nitrate in water, the segmented flow and non-segemented flow instrument schematics for which are illustrated in Figure 10.9. Both systems make use of the cadmium reduction method described earlier in the section on spectrophotometric analysis. As can be seen, in the segmented system, the sample and a solution of ammonium chloride* are combined in the air segmented flow stream and then passed through a short mixing coil. Exiting the mixing coil, the stream passes through the cadmium reduction column. Exiting the reduction column, the stream is mixed with the diazotization reagent (composed of sulfanilamide and N-(1-naphthyl)-ethylenediamine dihydrochloride). It is then delayed to allow the diazotization reaction to proceed by passing it through a longer mixing coil from which it exits to enter the colorimeter.

Except for the extra tubes providing the segmentation air, the arrangement of the non-segmented system is quite similar, however, with one important difference being the manner in which the sample is inserted into the system. Note that in the segmented system, sample is aspirated from its container directly into the flow system by means of the peristaltic pump. It is then merged with the air segmented ammonium chloride solution, mixing with it in the interspersed reaction chambers (Figure 10.9b). Because of the separation the air segments provide, the overall length of the sample zone in the flow stream is not critical. The response profile in the detector is a series of pulses of first increasing and then

*The ammonium chloride solution is a mildly acidic salt and provides the acidity required for the diazotization reaction to proceed.

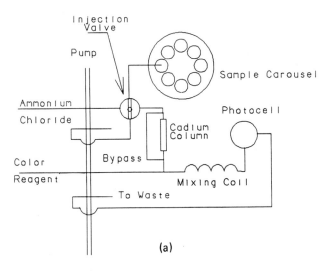

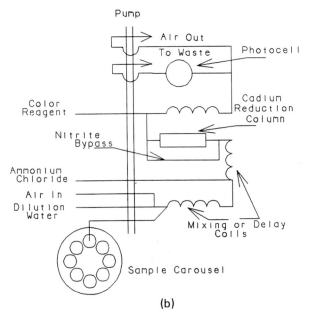

Figure 10.9 (a) A schematic of a flow injection analyzer for determination of nitrate or nitrite in acid rain. The by-pass is used for nitrite analysis. (b) A segmented flow analyzer for the determination of nitrate or nitrite in acid rain.

decreasing magnitude, corresponding to the distribution of the analyte in the segmented chambers.

In contrast, in non-segmented flow, the sample is aspirated through an injection valve to waste. The filled injection valve is rotated to inject the sample directly into the flowing stream, forming a sample zone that mixes progressively with the leading and trailing streams of ammonium chloride (Figure 10.9a). However, flow in non-segmented systems is always adjusted to be in the laminar region. That is, the flow velocity profile across a diameter of the flowtube is parabolic, with the center of the stream moving at about twice the stream's mean flow rate while the periphery near the tube walls moves at very low flow rates. Consequently, the analysate flows into the conversion reactant at the leading edge of its zone, and the conversion reactant flows into the analysate at the trailing edge. Mixing in the stream is never complete, but although never complete, it may be made to be very reproducible, allowing, as discussed above, quantitation of the analyte to be achieved.

Environmental Applications of Continuous Flow Analyzers

The analytical procedures used in continuous flow analyses are generally patterned after established manual analytical protocols. Thus, many of the methods that we have already discussed have been automated and have received EPA approval in the automated mode as well. The schematic shown in Figure 10.9 represents an EPA approved method for determination of both nitrate and nitrite in acid rain [Operation and Maintenance Manual 1986, Appendix H], the nitrite being determined by by-passing the cadmium reduction column.

Schematics of other typical systems are illustrated in Figures 10.10 and 10.11, in which EPA approved procedures for the determinations of sulfate and ammonia, respectively, in acid rains are illustrated [Operation and Maintenance Manual 1986, Appendices G & L]. The method adapted for determination of sulfate (Figure 10.10) is based on the decrease in concentration (and color intensity) of a complex that forms between barium ions and methylthymol blue dye as barium is removed from the system by the formation of barium sulfate. Barium chloride and methylthymol blue are mixed with the sulfate-containing sample stream, which is first passed through an ion exchange column to remove cations that interfere with the formation of the complex. The sodium hydroxide is added to the stream to intensify the color of the complex.

The indophenol blue method for ammonia illustrated in Figure 10.11 is an adaptation of the same method discussed earlier in the section on spectrophotometric measurements.

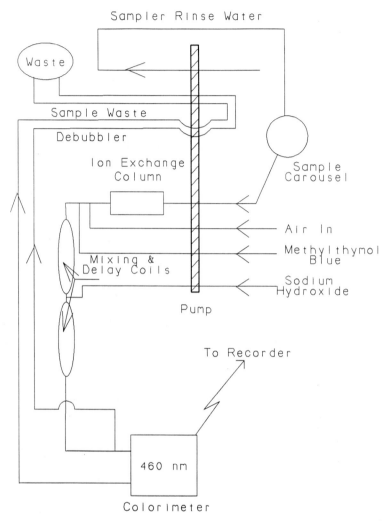

Figure 10.10   Schematic of a segmented flow analyzer for determination of sulfate.

It should be noted, incidentally, that as of this writing, EPA approval applied only to air segmented flow systems — not flow injection analyzers for analysis of acid rain. However, with the rapidly increasing utilization of flow injection analysis, it is a virtual certainty that regulatory agencies will ultimately classify it as universally equivalent to air segmented methods for all environmental contaminants.

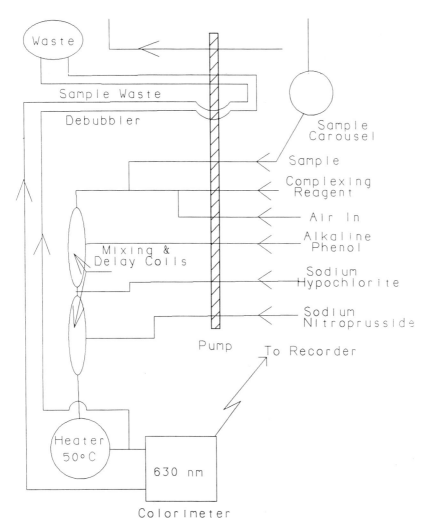

Figure 10.11   Schematic of a segmented flow analyzer for determination of ammonia.

Also, an interesting sidelight and a tribute to the broad utililization of continuous flow analysis is that it has generated a small industry of its own, devoted to the development and marketing of chemistry kits for AutoAnalyzers. These kits contain the appropriate reagents, tubing, and analysate treatment modules for the determination of hundreds of different analytes, among which are many that are applicable to environmental

monitoring. Typical of suppliers of such kits is the Skalar Corporation of
the Netherlands, which provides them for the determination, among
others, of sulfite, sulfate, nitrate, nitrite, carbonate, and carbon dioxide
in water.

A flow schematic for a Skalar carbonate determination is shown in
Figure 10.12. The protocol involves the acidification of the sample stream,
releasing carbon dioxide. The stream is then passed through a module
containing a gas permeable membrane, where the CO_2 diffuses into a
collector stream containing a mildly buffered solution of an acid-base

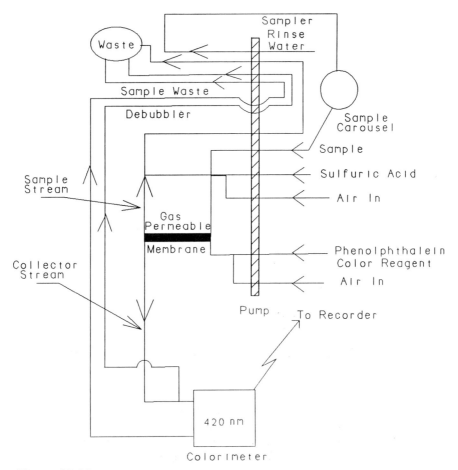

Figure 10.12   Schematic of a segmented flow analyzer for determination of
carbon dioxide in waste water.

indicator, the consequent pH and associated color change of which is proportional to the carbonate level in the sample.

Finally, it is also worth mentioning that as of the time of this writing, the Industrial Systems Division of the Technicon Corporation was in the process of developing AutoAnalyzer methods for determination of COD, TOC, and cyanide in water and waste water as well methods for determining Kjeldahl nitrogen in food, soil, and plant tissue extracts.

11

Bioanalytical Methods

INTRODUCTION

Bioanalytical methods, specifically enzymological and immunochemical methods, have been used for over a generation in clinical and forensic laboratories for the analysis of drugs and the components of physiological fluids. However, relative to most of the physical and chemical methods we have discussed thus far, their use for environmental analysis is a recent phenomenon. In fact, although no bioanalytical procedures had yet been approved for any pollutant by the EPA at the time of this writing, increasing numbers of environmental applications were beginning to appear in the technical literature, especially for the analysis of pesticides (Table 11.1) [Schalbe–Fehl 1986], and reports of developments in progress of tests for dioxin, benzene, and PCB's were beginning to be announced in trade publications [Krieger 1987; Shamel and Chow 1987]. Because of the potentially exquisite specificity and high sensitivity these methods can provide, as well as their procedural simplicity and relative economy, it seems that their recognition and approval by regulatory agencies must be inevitable. We are, therefore, including this discussion of the principles of enzymological and immunochemical analysis in anticipation of increasing utilization of the methods, especially for identification of specific contaminants in water and waste water.

ENZYMOLOGY AND ENZYME ANALYSIS

Enzymes

Enzymes are proteins that serve as catalysts for the thousands of chemical reactions that are always underway in living cells. They represent the

298

Table 11.1   Labeled Immunoassays for Some Pesticides[a]

Compound name	Use	Type of label[b]
Diflubenzuron	Insecticide	Enzyme
Parathion	Insecticide	RIA, enzme
Paraquat	Herbicide	RIA, enzyme
S-Bioallethrin	Insecticide	Enzyme
Dieldrin, aldrin	Insecticide	RIA
Benomyl and metabolites	Fungicide	RIA
2,4-D, 2,4,5,-T	Herbicide	Enzyme
Diclofop-methyl	Herbicide	Enzyme
Terbutryn	Herbicide	Enzyme
Metalaxyl	Fungicide	Enzyme
Chlorsulfuron	Herbicide	Enzyme
Triadimefon	Fungicide	Enzyme

[a]From [Schwalbe–Fehl 1985].
[b]Enzyme = Enzyme labeled assay. RIA = Radio labeled assay (see text).

largest and most highly specialized class of protein molecules in living matter, each of them catalyzing only a specific reaction type involving only very specific reactants, which are called the *enzyme substrates*. As catalysts, they are capable of increasing the velocity of their characteristic chemical reactions up to 10^8 fold over the uncatalyzed reactions. What this amounts to is that reactions that under ordinary circumstances appear not to proceed at all, proceed very rapidly in the presence of an appropriate enzyme. Over 2000 enzymes have been identified, many of which have been isolated and purified. And, as evidence of their proliferation in living matter, it has been estimated that there may be as many as 10^4 of them in a single bacterial cell [Robyt and White 1987, 291].

Enzymes are classified according to the types of reactions that they can catalyze. Some exhibit absolute specificity for a substrate while others act on a broader class of molecules having some common structural feature. Thus, there are enzymes that can distinguish and act specifically on one of a pair of sterically isomeric molecules. That is, they can distinguish differences as subtle as the geometrical orientation of a functional group on a substrate molecule. Other enzymes catalyze oxidation-reduction reactions and act on classes of chemical groups like alcohols, ketones, or amines. Others catalyze hydrolysis reactions, that is, they accelerate the splitting of functional groups like esters, peptide bonds, or glycosidic groups. And still others are capable of catalyzing the transfer of specific functional groups from one moiety to another, an example being an amine transferase, an enzyme that can transfer an amine group from an amino acid to an α-keto acid (see section on biochemistry). It is the specificity of enzyme catalyzed reactions that makes them so attractive as analytical reagents, since they are able to seek out their substrates in the presence of hosts of other substances, thus precluding necessity to carry out complicated separations to conduct analyses.

The reaction specificity exhibited by enzymes is thought to derive from an active site on the protein molecule that can geometrically accomodate the substrate molecule in a kind of lock and key configuration, with the active site serving as the lock. However, it has also been recognized that the lock and key model is something of an oversimplification. Thus, studies of enzyme specificity have indicated that two distinct features are required of the substrate. It must possess the particular chemical bond or functional group that can be attacked by the enzyme, and it must possess a second functional group that binds to the enzyme, positioning the reacting group properly on the active site so that it may be attacked. The binding entity on the substrate is also usually found to have some specific geometrical relationship to the entity that the enzyme attacks [Lehninger 1970, 170].

Historically, enzymes have been named by adding the suffix -*ase* to the name of the substrate. Thus, ure*ase* catalyzes the hydrolysis of urea to ammonia and carbon dioxide, and phosphat*ase* catalyzes the hydrolysis of phosphate esters. Unfortunately, the large diversity of enzymes and enzyme functions has made this nomenclature cumbersome; and non-systematic common names have also come into use like trypsin and catalase, the former of which catalyzes the hydrolysis of the peptide groups in proteins while the latter catalyzes the oxidation of certain organic dyes by peroxides. Consequently, an international classification and numbering system has been established, which is shown in Table 11.2. The international classification divides enzymes into six classes by function, each of which is then further divided into subclasses and sub-subclasses that designate the substrates upon which the enzymes act. In general, the systematic nomenclature of the international system is used in the technical literature when reference is made to enzyme function, but the more traditional names continue to be used in day-to-day references and in suppliers' catalogs.

Some enzymes require so-called *co-factors* to be present for them to exhibit catalytic activity. Co-factors are generally not proteins but may be metal ions or organic molecules. When the co-factor is an organic molecule, it is frequently referred to as a *coenzyme*. Co-factors may be bound to the protein moiety with varying degrees of strength up to and including covalent binding and, except for the covalently bound ones, they may be removed from the protein by a number of different means as, for

Table 11.2 International Classification of Enzymes

Class	Reaction type	Acting on (examples)
1. Oxido-reductases	Oxidation-reduction	1.1 Alcohols
		1.2 Ketones
		1.3 Double bonds
		1.4 Amines ($-NH_2$)
		1.5 Amines ($-NH-$)
		1.6 NADH, NADPH
2. Transferases	Functional group transfers	2.1 One-carbon groups
		2.2 Aldehydic, ketone groups
		2.3 Acyl groups
		2.4 Glycosyl groups
		2.7 Phosphate groups
		2.8 Sulfur groups
3. Hydrolases	Hydrolysis reactions	3.1 Esters
		3.2 Glysosidic bonds
		3.4 Peptide bonds
		3.5 Other C–N bonds
		3.6 Acid anydrides
4. Lyases	Addition to double bonds	4.1 Carbon–carbon bonds
		4.2 Carbon–oxygen bonds
		4.3 Carbon–nitrogen bonds
5. Isomerases	Isomerization reactions	5.1 Optical isomers
6. Ligases	Forms bonds with use of ATP	6.1 Carbon–oxygen bonds
		6.2 Carbon–sulfur bonds
		6.3 Carbon–nitrogen bonds
		6.4 Carbon–carbon bonds

example, dialysis. When the bonding is covalent and the co-factor cannot be separated from the protein, it is often called a *prosthetic group*. The intact enzyme-cofactor entity, regardless of the lability of the co-factor, is called a *holoenzyme* while the separated precursor protein of an enzyme-coenzyme pair is called an *apoenzyme*.

Elementary Enzyme Kinetics

Characteristic of enzyme catalyzed reactions is the fact that the rate at which substrate is converted to reaction product varies with the concentration of the substrate. The reaction velocity increases with concentration up to a maximum and then becomes constant. In 1913, a simple model was proposed to explain the kinetics of the substrate conversion [Michaelis and Menton 1913]. It is by application of that model and its resulting formulas that quantitative enzymatic analysis is conducted.

The Michaelis–Menton model is based on the assumption of a two stage reaction between the substrate and the enzyme to form the product. In the first stage, the enzyme and substrate form a complex (ES). In the second, the complex breaks down to produce the reaction product and regenerate free enzyme. The proposed equilibrium is expressed

$$E + S \underset{k_3}{\overset{k_1}{=}} ES \underset{k_4}{\overset{k_2}{=}} E + P \tag{11.1}$$

where E represents the enzyme, P the product, and the k's represent reaction rate constants which have the following meaning.

In a chemical reaction in which a single reactant is converted to a single product, say

$$X = Y$$

the rate of disappearance of X will be proportional to its concentration, and the rate of appearance of Y will be equal to the negative of the disappearance of X. That is, one may write for the rate of the conversion reaction

$$-\frac{d[X]}{dt} = \frac{d[Y]}{dt} = k[X]$$

where k, the proportionality factor, is called the reaction rate constant.

Similarly, in a reaction involving two reactants,

$$X + Z = Y$$

the rate of transformation to product can depend upon the concentrations of both, that is

$$\frac{d[Y]}{dt} = k[X][Z]$$

again with k representing the proportionality or reaction rate constant.

Thus, with these definitions in mind, the constants k_1 and k_2 are defined as the reaction rate constants for the forward reactions of the first and second stages of eq. 11.1, and k_3 and k_4 are the rate constants for the reverse reactions. Consequently, the rate of formation of the enzyme-substrate complex, ES, may be written

$$\frac{d[ES]}{dt} = k_1[E][S] \tag{11.2}$$

But since the concentration of free enzyme must be equal at any time to the total amount of enzyme in the system less the amount that is bound in the enzyme-substrate complex, one can also write eq. 11.2 as

$$\frac{[ES]}{dt} = k_1[(E_t) - (ES)][S]$$

Enzyme-substrate complex is also used up in the reactions of eq. 11.1, its disappearance depending upon the rate of the reverse reaction of the first stage and the rate of the forward reaction of the second stage. Thus, the disappearance of ES in the overall reaction may be expressed by the relationship

$$\frac{-d[ES]}{dt} = k_3[ES] + k_2[ES]$$

But after a finite reaction time, if sufficient reactant and enzyme are present, enzyme reactions achieve a steady state. That is, the rate of production of product neither accelerates or decelerates, and the concentration of ES remains constant. At steady state, therefore, the rate of formation of ES and its disappearance must be equal, and we have

$$k_1[(E_t)^- (ES)][S] = k_3[ES] + k_2[ES] \tag{11.3}$$

Then, lumping the rate constants of eq. 11.3 and re-arranging, one gets

$$[ES] = \frac{[E_t][S]}{K_m + [S]} \tag{11.4}$$

where K_m is called the Michaelis constant and is characteristic of a particular enzyme-substrate combination.

But the overall conversion rate of S to P depends upon the concentration of ES at steady state. That is, the rate of production of P is given by

$$\frac{d[P]}{dt} = V_r = k_3[ES] \tag{11.5}$$

and substituting for [ES] in eq. 11.4 yields

$$V_r = k_3 \frac{[E_t][S]}{K_m + [S]} \tag{11.6}$$

Finally, in enzymatically catalyzed reactions, the substrate concentration is normally very much higher than that of the enzyme. So the equilibrium of eq. 11.1 will be such to drive the reaction to the right, converting essentially all the enzyme to enzyme-substrate complex. Under the condition that [ES] is equal to [Et], the maximum possible production rate of P is achieved, and one may write, using eq. 11.5

$$\left(\frac{dP}{dt}\right)_{max} = V_{max} = k_3[E_t] \tag{11.7}$$

Then, dividing eq. 11.6 by 11.7, one arrives at the final form of the Michaelis–Menton equation

$$V_r = \frac{V_{max}[S]}{K_m + [S]} \tag{11.8}$$

which, as can be seen, predicts the observed behavior of enzyme catalyzed reactions. That is, at low values of [S], where $[S] \ll K_m$, the reaction velocity is proportional to [S] while at high substrate concentrations, where [S] is $\gg K_m$, the velocity approaches a constant value that is independent of the substrate concentration.

Quantitative Analysis via Enzyme Reactions

Equation 11.8 may be used directly to determine the concentration of an unknown substrate in an analysate. By fixing the enzyme concentration and, thus, the value of V_{max}, and measuring the reaction velocity as a function of substrate concentration, one can develop a calibration curve, which may then be used to determine the levels of substrate in unknowns.

The formal protocol of an enzyme procedure, however, generally involves first measuring the value of K_m for a given enzyme so as to be able to determine the range of substrate concentrations that will exhibit linearity with reaction velocity. Working in the linear region of eq. 11.8 precludes necessity for the formalism of a calibration curve since, in the linear region, responses will be directly proportional to [S], and one need only measure the response of an unknown and relate it to that of a standard to determine the value of the unknown.

Access to the value of K_m is through a re-arranged form of the Michaelis–Menton equation. That is, one takes the recipocal of eq. 11.8, the result of which is called the Lineweaver–Burke equation

$$\frac{1}{V_r} = \frac{K_m}{V_{max}[S]} + \frac{1}{V_{max}}$$

Then, by measuring V_r as a function of [S] and plotting their reciprocals against one another, one can calculate K_m from the values of the slope and intercept of the linear Lineweaver–Burke curve.

Generally, as will be discussed below, reagents for enzymatic analysis are provided as kits with the value of K_m for the enzyme involved having been previously determined by the supplier, and the range of linearity over which the kit may be applicable being specified in the kit's instructions.

Enzyme analyses are almost always based on some form of colorimetry or spectrophotometry, measuring the rate of change of absorbance resulting from either the disappearance of the substrate or the appearance of the product. Sometimes, however, the reaction product and the substrate of analytical interest exhibit similar light absorption characteristics with little or no change in absorbance resulting from the reaction. In those cases, cascaded enzyme reactions may be used in which the product of a first reaction serves as the substrate of a second reaction where there is a net change in absorbance.

An example of such an analysis is the enzymatic determination of glucose. Glucose is oxidized by the enzyme glucose oxidase yielding two products, glucuronic acid and hydrogen peroxide, with no change in absorbance generated by the reaction. So, advantage is taken of the oxidation of an indicator dye by the hydrogen peroxide generated, which is catalyzed by the enzyme horseradish peroxidase. A large excess of the peroxidase is added to the system so that the indicator oxidation occurs as fast as the hydrogen peroxide is generated, making the rate of change in color of the dye exactly equal to the reaction velocity involved in the production of the hydrogen peroxide and, thus, the oxidation of the glucose.

Typically, reaction velocities are determined by simply allowing the enzyme(s)-substrate reaction(s) to proceed for a fixed time at a fixed temperature in some appropriate reaction vessel.* The change in absorbance in the reaction mixture is then measured in a colorimeter or spectrophotometer and related to the substrate concentration.

---

*The term "incubate" is frequently applied to this operation by bioanalytical chemists.

Environmental Applications of Enzymatic Analysis

Commercial enzyme kits are available that may be applied to determination of ammonia, sulfites and nitrates in water.* The kits generally contain all the necessary reagents to carry out the analyses along with instructions and dilution formulas that preclude necessity to pre-determine factors like K_m and V_{max}. Brief descriptions of the chemistry involved in the analyses follow.

Ammonia

Determination of ammonia is based on its reaction with 2-oxo-glutarate in the presence of the enzyme glutamate dehydrogenase and its coenzyme NADH

$$2\text{-oxoglutarate} + NADH + NH_4^+$$
$$\xrightarrow[\text{glutamate dehydrogenase}]{} L\text{-glutamate} + NAD^+ + H_2O$$

The rate of NADH consumption is monitored via its absorption at 340 nm in a uv spectrophotometer.

The system may be used for determination of ammonia in water or for the end determination of ammonia in a Kjeldhal nitrogen analysis. The sensitivity of the determination is of the order of 20 ppm.

Sulfites

Sulfite is determined by means of two enzymes, sulfite oxidase and NADH-peroxidase, an enzyme that oxidizes NADH to NAD^+ in the presence of peroxide. The reactions are

$$SO_3^{2-} + O_2 + H_2O \xrightarrow{\text{Sulfite Oxidase}} SO_4^{2-} + H_2O_2$$

$$H_2O_2 + NADH + H^+ \xrightarrow{\text{NADH-peroxidase}} 2H_2O + NAD^+$$

with the NADH being measured spectrophotometrically. As in the glucose determination described above,[†] the NADH-peroxidase and NADH

---

*Boehringer–Mannheim Corporation, Indianapolis, Indiana.
[†]One might wonder why the manufacturer chose to use NADH-peroxidase in preference to horseradish peroxidase, as is done in the determination of glucose. The reason is that sulfite binds to horseradish peroxidase, destroying its catalytic activity.

are added in sufficient quantity to maintain the oxidation of the sulfite the rate-limiting reaction. The sensitivity of this determination is claimed to be of the order of 30 ppm of sulfite. The system may be used to measure sulfite in water and waste water among other things.

Nitrates

In this determination, nitrate is reacted with NADPH in the presence of the enzyme nitrate reductase. The NADPH is thus oxidized to $NADP^+$, its disappearance being measured spectrophotometrically at 340 nm. The reaction is

$$NO_3^- + NADPH + H^+ \xrightarrow{\text{nitrate reductase}} NO_2^- + NADP^+ + H_2O$$

The sensitivity of the measurement is of the order of 30 ppm of nitrate, and the system may be used to determine the nitrate levels in water and waste water.

Disadvantages of Enzyme Analysis

Proteins, in general, are fragile structures. They are large, complex molecules whose properties, especially their enzymatic properties, are strongly dependent upon their geometric configuration. This complexity and fragility leads, therefore, to the most serious disadvantage of enzymatic analysis. That is, enzymes have a high proclivity for de-activation. Elevated temperatures and sometimes just mild fluctuations in pH can denature them, i.e., change their physical configuration, and destroy their activity. Enzymes are also easily poisoned by some metal ions, and they are subject to competitive inhibition, a situation in which their active sites are blocked by non-substrate substances. As a consequence, their utility for analyses in some waste waters may be limited.

IMMUNOCHEMISTRY AND IMMUNOCHEMICAL ANALYSIS

The Immunological Response, Antibodies, and Antigens

Higher animals have a uniquely protective response to the entry of certain foreign substances into their systems. They produce *antibodies* — a class of proteins that chemically bind the substance so as to neutralize it and minimize any physiological damage it might incur. Substances that can elicit the production of antibodies are called *antigens*, and antigens and antibodies share a very important property. They react with one another but not with other substances. That is, an antibody can bind only to the antigen that elicited its production, and an antigen can bind only to the antibody it elicited. It is this extraordinary specificity that makes antigens and antibodies (*immunochemicals*) so attractive as reagents for analytical procedures.

Antibodies are produced in the blood stream of an animal by repeatedly injecting it with antigen over a period of time. The repeated injections stimulate the animal to produce a high concentration (or *titre*) of antibody, which is then harvested by drawing aliquots of the animal's blood. Because of the high specifity antibodies exhibit, using them generally requires little more purification than centrifuging the harvested blood to remove its red cells. The antibody-containing serum, i.e., the supernatant fluid remaining after the red blood cells are removed, is generally referred to as *antiserum*.

For a foreign substance to be recognized as such by an animal, i.e., to be antigenic, it must be a large molecule with a molecular weight in excess of about 1000. However, antibodies to smaller molecules may be produced by injecting an animal with a chemical combination of the small molecule and a large one that is, itself, antigenic. Thus, one could produce antibodies to an insecticide, for example, by chemically coupling the insecticide to a protein molecule and using the couple as the antigen. Injection of such a *conjugated antigen* into an animal would elicit antiserum that contained two kinds of antibodies, however, one that was anti- the protein and one that was anti- the insecticide. But in using such an antiserum for determination of the insecticide, it would be unnecessary to separate the two types of antibodies, as long as the unknown material did not contain the protein to which the insecticide had been coupled originally.

Antibodies to small molecules that, by themselves, are not antigenic are said to be *hapten-specific*; and the small molecules themselves, those that elicit antibody responses as part of conjugated antigens are called *haptens*.

Antibodies, being proteins, may also serve as antigens in other animals, producing anti-(antibodies). Thus, injection of an antibody to a human protein raised in a rabbit into a goat, for example, would elicit a response in the goat, producing goat anti-(rabbit antibody). The goat anti-(rabbit antibody) would then be reactive toward the rabbit antibody that elicited it but not toward the human protein that was injected into the rabbit. That is, in general, anti-antibodies are not reactive toward the antigens that produced the original antibodies.

Antibody and Antigenic Binding Sites

Antibody proteins are part of a general class of proteins called immunoglobulins that have the characteristic 'Y' shaped molecular structure shown in Figure 11.1a. The tines of the molecule are called the F_{ab} fragments, and the stem of the molecule is called the F_c fragment. Each of the F_{ab} fragments contains a recognition and antigen binding site. Thus, each antibody molecule may bind two antigen molecules. Antigens, on the other hand, may be monovalent, i.e., have only a single binding site, or multivalent, i.e., have many binding sites.

When monovalent antigens react with antibodies, the resultant Ab–Ag* complexes will generally remain in solution, each one containing

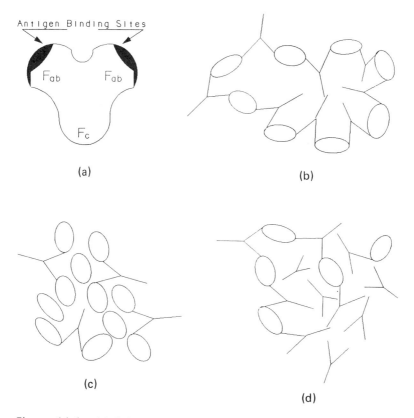

Figure 11.1 (a) Schematic of an antibody molecule. The antigen recognition and binding sites reside on the F_{ab} fragments of the molecule. (b) array of antibody and antigen molecules at their equivalence point. (c) Binding antigen is in excess. (d) Binding when antibody is in excess.

*This is the shorthand notation for an immune complex.

one or two bound antigen molecules, depending upon the concentration ratios of the antibodies and antigens present.

When multivalent antigens and the bivalent antibodies react, however, there are several results possible that depend upon the concentration ratios of the reacting entities. When the antibodies and antigens are in molar equivalent ratios, the resulting Ab–Ag complex can form the array structure shown in Figure 11.1b. That is, each multivalent antigen molecule may serve as a bridge between the binding sites of neighboring antibody molecules. In such cases, the complex can precipitate (form a *precipitin*). If there is too large an excess of antigen molecules, however, the binding sites of the antibody molecules will be saturated by individual antigen molecules, and no precipitin will form (Figure 11.1c). On the other hand, if there is an excess of antibody, the formation of precipitin will be limited by the amount of antigen available to bridge between antibody molecules (Figure 11.1d).

The binding of antigen to antibody is a reversible phenomenon, whether the antigen is mono- or multivalent. Thus, for example, both the hapten and the hapten-antigen conjugate used to elicit an antibody would compete for antibody binding sites in a reacting system, and the ratio of the bound entities could be varied by changing their concentration ratios in the system. Thus, if one first reacted the antibody with the conjugate alone and saturated the binding sites, one could then displace the conjugate by adding an excess of hapten alone.

In the case of multivalent antigens where a precipitin has formed, if excess antigen is added, reversibility allows it to substitute into the bridging antibody-antigen bonds and re-solubilize the precipitin by saturating the antibody binding sites. Consequently, given a solution containing some concentration of antibodies and titrating it with varying quantities of antigen molecules, one obtains a curve of the kind shown in Figure 11.2.* As can be seen, the amount of precipitin formed is at maximum when the the concentrations of antibody and antigen are in molar equivalence. With excess antigen, the precipitin re-dissolves; with excess antibody, the amount of precipitin formed depends upon the amount of antigen added.

---

*Titration in the context of immuno-reactions means reacting individual aliquots of varying ratios of antibody and antigen in separate vessels. The individual vessels are then incubated for a time sufficient to allow the precipitin reaction to go to completion, and the amount formed in each vessel is determined. It is not meant to imply a titration of the kind discussed earlier in the section on volumetric analysis.

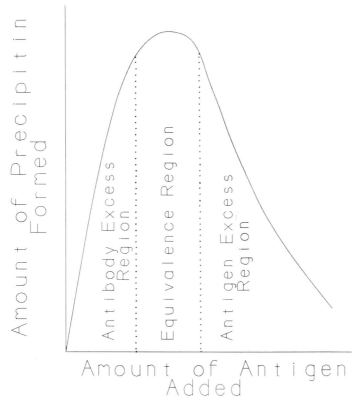

Figure 11.2   An antibody-antigen titration curve illustrating the behavior of the precipitin formed as antigen is added to antibody.

Analytical Applications of Immunochemicals

Detailed discussion of the vast array of adaptations of immunochemicals to quantitative and qualitative analysis is well beyond the scope of this monograph. Suffice to say here that all of the methods are based on first forming an immune complex, with some analyses based simply on the formation of precipitins and others based on the physical separation of the complex from the reaction mixture.

Because of the specificity of the reactions, direct identification of antigens (or antibodies) may be made from the formation of precipitins, and many qualitative immunochemical analyses are based on that principle [Freifelder 1982, 326; Hudson and Hay 1980, Chapter 5; Morris, Smith, and Greyson, 1976]. Small molecule (hapten) identifications may be also be made through their ability to inhibit precipitin formation or dissolve them, i.e., through their ability to compete with or displace the large multivalent hapten-conjugate antigens that form the precipitins. Historically, qualitative immunochemical analyses based on precipitin formation were among the first developed and have played a very important role in the identification of antigens and antibodies in the diagnosis of disease.

But most important to environmental analysis are those more recently developed immunochemical methods that are based on the use of labeled antigens or antibodies to quantitate analytes. As we shall see, radioactive and enzyme labels on antigens or antibodies provide a route to analyte determinations with a degree of sensitivity and specificity rivaled by few, if any, other analytical approaches.

Radioimmunoassays

The radioimmunoassay (RIA), i.e., the use of radiotracer labeled immunochemicals for quantititive analysis, was developed in the late fifties by Berson and Yallow [Berson and Yallow 1959; Yallow and Berson 1968].* The principle of the method is straightforward. Antibody is combined with labeled antigen. Adding unlabeled antigen to the system will then displace the labeled antigen. By measuring the radioactivity of the displaced labeled antigen or the residual radioactivity on the separated Ab–Ag complex, one has a measure of the amount of unlabeled antigen that was added to the system.

Practical application of an RIA requires the development of a calibration curve first. This is done by placing a fixed ratio of labeled antigen (Ag*) and antibody into a series of reaction tubes. Known quantities of unlabeled antigen are then added to each of tubes. When the reactions are complete, the immune complex in each tube is separated from the mixture, and either its activity or the activity remaining in the tube is measured. If the activity of each of the separated complexes from each tube is measured, the values obtained will be inversely proportional to the amount of unlabeled antigen added to each tube. If the residual activity in each of the tubes is measured, the values will be directly proportional to the amount of unlabeled antigen added to each. A calibration curve relating the quantity of added unlabeled antigen to the amount of radioactivity measured is then constructed.

To conduct an analysis, an aliquot of an unknown sample is added to

---

*It is worth noting that the import of the radioimmunoassay, in medical applications at least, was recognized by awarding the Nobel prize to Berson and Yallow for its development.

a tube containing the same mixture of Ab–Ag* that was used to develop the calibration curve. And, as before, after completion of the reaction, the complex is separated from the mixture. The measured activity, either residing on the complex or residual in the tube, can then be used to pick the concentration of the antigen in the unknown off the appropriate calibration curve.

There are a number of ways to separate the immune complex from the reaction mixture. One of the most frequently used, when analyzing small molecules, is to adsorb the small molecules on activated charcoal that has been coated with Dextran.* The Dextran serves as a filter, excluding the large Ab–Ag complexes from the charcoal surface. Thus, the small molecules selectively adsorb on the charcoal. Centrifugation of the system then separates the small molecules from the immune complex, which remains in the supernatant fluid.

An alternative separation method makes use of a second antibody that is anti- to the antibody of the assay. Addition of an equivalent quantity of the second antibody to the system will then form a precipitin with the immune complex, which may also be separated from the mixture by centrifugation.

The sensitivity of RIA'S is remarkable, analyte concentrations as low as picomolar (10^{-12} moles/liter) being measurable for some systems. But their are disadvantages involved as well. First, the assays require the use of radioactive materials, which can be hazardous and which generally require analyst licensing. Secondly, although the assay manipulations are relatively straightforward, the equipment for measuring radioactivity can be expensive; however, no more so than some of the sophisticated instruments that have been introduced earlier. Perhaps the most serious disadvantage is the assay time. Reaction times, that is the formation of the immune complexes, especially for analysis of very low levels of analytes, can be protracted, as long as one or two days. Thus, in situations where turn-around time in analysis is critical, RIA's may sometimes be inappropriate. Additionally, although a less serious problem, is that radiolabeling sometimes can diminish or even destroy the activity of immunoreactive reagents. Thus, the method is not necessarily universal in its applicability.

Enzyme Labeled Immunoassays

The general principle underlying all enzyme labeled immunoassays is the linking of an enzyme capable of generating a measurable reaction product

---

*Dextran is a polysaccharide based molecular sieve supplied by the Sigma Chemical Co., St. Louis, Mo.

to one of the reacting species of a test immume reaction. Similar to RIA's, quantitation is achieved by measuring the enzyme activity associated with the isolated immune complex or the residual enzyme activity after the immune complex is separated from the test system. Extraordinary sensitivity, comparable to that of RIA's, may be achieved because of the amplification factor that the enzyme-substrate reaction permits. That is, one may use the enzyme reaction, in theory, to provide as much response as necessary to achieve whatever sensitivity is desired by simply adding sufficient substrate and allowing sufficient time for the reaction to proceed. Even picomolar quantities of labeled analyte may thus be detected, depending upon the reaction system.

Conceptually, the simplest of enzyme labeled assays might make use of an enzyme labeled antigen to compete with the analyte antigen for binding sites on an antibody — just as in RIA's. Separation of the immune complex could be effected by the formation of a precipitin with an anti-antibody as well. And generation of a calibration curve for the system would also be identical to that of an RIA determination, i.e., by means of a series of tubes containing known ratios of unlabeled and labeled antigen.

However, in general, enzyme labeled immunoassays are carried out in more heterogeneous systems. That is, one of the reacting immuno-reagents is first bound to a solid support, frequently by simple physical adsorption. Such analytical systems are known as *enzyme labeled immuno-sorbent assays*, or *ELISA*.

In some ELISA protocols, the counter species and a known quantity of its labeled analog are combined with the sorbed immuno-reagent and allowed to compete for binding sites on it, after which the solid phase is washed free of the reaction mixture and its enzyme activity measured. In such systems, the measured activity is inversely proportional to the amount of unknown present. That is, larger amounts of unknown compete more effectively for the available binding sites, resulting in fewer bound labeled molecules.

In other protocols, the total binding capacity of the sorbed immuno-reagent is calibrated by means of a labeled counter species, which may be a labeled antigen or a labeled anti-antibody in the case of a sorbed antibody. Unknown is then combined with a known quantity of the calibrated solid material. Following washing the solid free of the excess analysate, it is then reacted with the labeled counter species, separated again from the reaction mixture and washed again. Enzyme substrate is then added to it and its enzyme activity is measured. As in the previous example, the amount of enzyme activity measured will be inversely proportional to the quantity of unknown in the analysate, because the higher the concentration of analyte, the fewer will be the available binding sites for the labeled material.

How the particular protocol for an ELISA assay is designed depends in large measure upon the system being analyzed. For the determination of a multivalent antigen, for example, one might first sorb the antigen's antibody to the solid support. The sorbed antibody would then be reacted with the unknown analyte. After washing the sorbed Ab–Ag complex, it would be reacted with enzmye labeled antibody, which could bind to the remaining available sites on the bound multivalent antigen. As before, measurement would follow separation and washing. In this case, however, the enzyme activity exhibited by the separated solid support would be directly proportional to the quantity of analyte in the analysate because it is that entity to which the labeled antibody would bind.

A monovalent antigen is treated somewhat differently. One method, similar to that described above in which the analyte is simply allowed to compete with an enzyme labeled analog, has been described for the determination of the herbicide terbutryn [Huber and Hock 1985]. The procedure entails first adsorbing herbicide antibody on the walls of the wells of a microtiter plate.* Samples and standards are then incubated in the wells. Following the incubation, the wells are washed and a complex of terbutryn coupled to the enzyme alkaline phosphatase is added. The antigen-coupled enzyme then binds to the remaining unbound sites of the antibody. After washing the microtiter wells again, the enzyme substrate is added, and the color produced from the enzyme reaction is measured.*

An alternative method is to coat a solid support with antigen. Sample antigen is then incubated with excess antibody in a separate vessel. When the incubated mixture is applied to the solid support, only the unreacted antibody, the quantity of which is inversely proportional to the sample antigen, will bind to the sorbed antigen. The sorbed immunocomplex is then washed and an enzyme labeled anti-antibody is added. After washing

---

*A microtiter plate is a plastic or glass plate with an array of small reaction wells cast in it. It is a convenient device to use for immunochemical analyses because the reactions underway in each of the wells may be readily compared to one another on a side-by-side basis. Additionally, other devices are available that allow many of the filling and washing operations as well as the spectrophotometric measurements associated with analyses in microtiter plates to be automated.

*Alkaline phosphatase catalyzes the decomposition of paranitrophenyl phosphate, which is colorless, to form p-nitrophenol. The latter, under alkaline conditions, is an intense yellow. The intensity of the color developed in the microtiter wells is directly proportional to amount of labeled antigen bound and inversely proportional to the amount of terbutryn in the samples.

again, the enzyme substrate is added to the system, and rate of product generation is measured. This technique has been applied to measurement of several insecticides [Schwalbe–Fehl 1986].

Regardless of the protocol used in an ELISA assay, it is alway necessary to prepare standard curves for an analysis. Additionally, immunogenic activity is highly variable; one batch of antibody may exhibit a very different titre than the next. Immunogenic activity is also subject to change with time. Thus, it is always necessary to prepare new standard curves for each new lot of material used as well as for materials that may have been stored for extended periods.

Environmental Applications of Immunoassays

Although they are mostly confined to research applications, a number of RIA and ELISA assays have been reported for insecticides, fungicides, and herbicides. Those listed in Table 11.1 may illustrate the potential of the technique for environmental analysis. In general, the assays reported have proved to be at least comparable to and often more sensitive than the chromatographic and mass spectrometric methods traditionally used for these substances.

Advantages and Disadvantages of Immunoassays

The most significant advantages of immunoassays are, of course, their exquisite specificity and their potential for extraordinary sensitivity. Relative to GC–MS, which might be regarded as their most significant competitor for the measurement of organic pollutants in water, the laboratory requirements of immunoassays are inexpensive and very unsophisticated; and the level of technical sophistication required to interpret the results of immunoassays is not nearly so high as that necessary to interpret the significance of chromatograms and mass spectrograms. Their major disadvantage is the necessity to raise specific antibodies to each of the pollutants to be determined. Raising antibody can be expensive, requiring animal maintenance facilities and highly specialized personnel familiar with antigen innoculation protocols. Thus, despite their very significant advantages, immuno-reagents may find practical application for the analysis of only the most ubiquitous pollutants.

Part IV
Commercial Equipment

12

Some Comments on Commercial Equipment

As a concluding chapter to this monograph, it seemed appropriate to include some discussion about commercial sources of apparatus and supplies, of which there are very many worldwide. These suppliers provide, in one form or another, all the devices and technology that we have been discussing. But additionally, because the world of analytical instrument manufacture is dynamic and competitive, most of the more reliable suppliers continually upgrade their offerings. Thus, there is not only a large array of available possibilities, but an array that is constantly changing.

Discussion of specific instruments or comparisons among the many that suppliers provide, therefore, becomes not only a potentially enormous undertaking, but one of questionable value because of timeliness, i.e., such a discussion would probably be seriously outdated by the time of publication of this monograph. Such a discussion would also serve only limited purpose in view of the specification of reference and equivalent methods by EPA, i.e., radical departures from approved methods would have, at best, only limited utility for regulated applications. Consequently, instead of attempting to provide specifics about suppliers and their products, it seemed more appropriate in these comments to indicate how suppliers may be identified and associated with the various technologies that we have been discussing. Table 12.1, therefore, is a list of supplier directories that are published by several different trade and technical magazines. These directories are revised annually or biannually. They provide indices by instrument category and cross reference them to manufacturers and distributors. Thus, for any specific analytical requirement, one need simply search the directories to identify potential suppliers.

Table 12.1 Directories of Instrument and Laboratory Chemical Suppliers

Name	Publisher	Address
American Laboratory Buyers Guide Edition	International Scientific Communications, Inc.	30 Controls Dr. P.O. Box 870 Shelton, CT 06484
Analytical Chemistry Labguide	American Chemical Society	1155 16th St. NW Washington, DC 20036
Chemcyclopedia	American Chemistry Society	1155 16th St. NW Washington, DC 20036
Guide to Scientific Instruments	American Association for the Advancement of Science	1515 Mass. Ave. NW Washington, DC 200054
ISA Directory of Instrumentation	Instrument Society of America	P.O. Box 1277 Research Triangle Park, NC 27709
Pollution Engineering Environmental Control Telephone Directory	Pudvan Publishing Co.	1935 Shermer Rd. Northbrook, IL 60062
Pollution Equipment News Catalog & Buyers Guide	Rimbach Publishing, Inc.	8650 Babcock Blvd. Pittsburgh, PA 15237
Research & Development Telephone Directory	Cahner's Publishing Co.	275 Washington St. Newton, MA 02158

In selecting suppliers for any particular application, however, consideration should be given to the distinction between those that supply *analyzers* from those that provide *monitors*. *Analyzers*, in general, are designed to analyze discrete samples extracted from a sample population while *monitors* are designed to analyze the sample population on a continuous basis. Although they may be used to determine the same analyte, the former is usually located remote from the sample population while the latter is designed to be an online device, equipped with a probe of some kind that resides within the sample population and senses changes in its character as a function of time. Thus, under the indexed category for sulfur analyzers, for example, one may find suppliers who provide combustion analyzers for use in a laboratory environment and suppliers who provide fluorescence analyzers that are designed to measure the SO_2 effluent from a smoke stack.

The distinction between analyzers and monitors is not intended to imply that these two devices may not be adapted to the functions of the other. They may and occasionally are. Gas chromatographs, which ordinarily would be categorized as laboratory analyzers, may, for example, frequently serve as monitors. They are equipped with automatic sampling valves that are inserted in the process stream, which are then used to extract samples periodically from the stream to inject them into their columns. There is also certainly no reason why an instrument used to measure the fluorescence of a sulfur dioxide effluent from a smoke stack could not be adapted to determine the sulfur dioxide content of discrete samples in the laboratory. But, in general, analysis instruments are best confined to the laboratory while monitors are best applied to on-line applications, if for no other reasons than analyzers tend to be more fragile than monitors and monitors tend to be less analytically sensitive than analyzers, being used primarily to sense changes in sample population rather than to determine absolute values of analytes; and suppliers that specialize in one or the other types of devices tend to have their expertise and their support capabilities centered around their specialty.

In addition to analyzers and monitors, suppliers in any specific analytical category may also specialize in devices that are distinguished as *personal monitors* or *field instruments*. Personal monitors are designed to be worn or carried by an individual, to detect a specific hazard in his environment or to determine the overall exposure to the hazard to which he may have been subjected. A film badge that time-integrates exposure to radioactivity is an example of a personal monitoring device. However, personal monitors may be much more sophisticated than a film badge as, for example, portable detectors for combustible hydrocarbon gases or carbon monoxide.

Field instruments are essentially laboratory instruments that have been portablized and ruggidized for use at sampling locations. Field instruments and laboratory instruments bear a relationship to one another that is similar to the relationship of personal monitors to process monitors. That is, personal monitors are moved through a sampling environment while process monitors are fixed at a sampling station. Field instruments are also moved through the sampling environment while the laboratory device is fixed in location. As above, field instruments may also be used in the laboratory and some personal monitors may be adapted to serve as process monitors. But as above also, it is best to use a device for its intended application, since most are designed for optimum operation in their own spheres.

Analytical devices may also be categorized as general or special purpose. Thus, ir, uv, and visible spectrophotometers, chromatographs, and mass spectrometers are all general purpose instruments, adaptable to determination of a wide variety of analytes taken from an equally wide variety of sources. On the other hand, a combustion analyzer designed to determine the carbon, nitrogen, and hydrogen composition of organic compounds is a moderately specialized device while one designed specifically for TOC determinations is clearly highly specialized as is the on-line monitor used to measure the SO_2 generated by a power plant. Laboratory instruments clearly include both general and special purpose devices. Monitors, however, are almost always highly specialized.

A general rule of thumb, although certainly not a hard and fast one, is that the more general purpose a device is, the more likely it is that it will require a higher level of technical sophistication to maintain it as well as to interpret the data that it generates. Thus, if a particular application lends itself to a special purpose device, it is probably wiser to obtain it in preference to a more general purpose instrument unless some anticipated future need dictates otherwise.

As a final point, in investigating supply sources for any particular application, consideration should always be given to the possibility of "letting someone else do it." That is, depending upon the anticipated analysis frequency, the apparent level of sophistication necessary to conduct the analyses, the personnel available to conduct them, and the cost of the required equipment, one may find it far more economical to contract the analytical activity to an external laboratory. An excellent reference for locating appropriate resources resides in the ASTM's* "Directory of Testing Laboratories," a listing of private and institutional analytical laboratories along with their analytical specialties.

---

*ASTM, 1916 Race St., Philadelphia, PA, 19103.

Bibliography

"Acid Rain and Transported Air Pollutants: Implications for Public Policy," Report OTA-0-204, U.S. Congress Office of Technology Assessment, Washington, D.C., June, 1984.

"Acid Rain — An EPA Journal Special Supplement," Report OPA-86-009, U.S. Environmental Protection Agency, Office of Public Affairs, Washington, D.C., September, 1986.

"Environmental Quality — 1984, 15th Annual Report of the Council on Environmental Quality," U.S. Government Printing Office, Washington, D.C., 1984.

"Environmental Quality — 1981, 12th Annual Report of the Council on Environmental Quality," U.S. Government Printing Office, Washington, D.C., 1981.

"Environmental Progress and Challenges: An EPA Perspective," Report PM-222, U.S. Environmental Protection Agency Office of Management Systems and Evaluation, Washington, D.C., June, 1984.

"Environmental Quality, 16th Annual Report of the Council on Environmental Quality," U.S. Government Printing Office, Washington, D.C., 1985.

Laboratory Robotics Handbook, Zymark Corp., Hopkinton, MA, 1988.

Methods in Microanalysis, Vols. I–VI, ed. Kuck, J. A., Gordon and Breach, New York, 1978.

"Operations and Maintenance Manual for Precipitation Measurement Systems," EPA/600/4-82-042b, U.S. Environmental Protection Agency, Center For Environmental Research Information, Cincinnati, OH, 1986.

"Quality Assurance Handbook for Air Pollution Measurement Systems," *Volume I, Principles*, EPA-600/9-76-005, U.S. Environmental Protection Agency, Research Triangle Park, North Carolina 27711, 1976; *Volume II, Ambient Air*

Specific Methods, EPA-600/4-77-027a, 1977; *Volume III, Stationary Source Specific Methods*, EPA-600/4-77-027b, 1977.

Standard Methods for the Examination of Water and Wastewater, American Public Health Association, Washington, D.C., 15ed., 1981.

40 CFR, "Code of Federal Regulations — Title 40," U.S. Government Printing Office, Washington, 1987.

Alberts, B., Bray, D., Lewis, J., Raff, M., Roberts, K., and Watson, J., *Molecular Biology of The Cell*, Garland, New York, 1983.

ASTM, *Annual Book of ASTM Standards*, The American Society of Testing and Materials, Philadelphia, PA, 1987.

Baird, D. C., *Experimentation: An Introduction to Measurement Theory and Experiment Design*, Prentice-Hall, Englewood Cliffs, New Jersey, 1964.

Bance, S., *Handbook of Practical Organic Microanalysis*, Wiley, New York, 1980.

Barnett, E. and Wilson, C. L., *Inorganic Chemistry*, Longmans, Green, & Co., London, 1959.

Battan, L. J., *The Unclean Sky*, Doubleday, New York, 1966.

Bernard, B. B., "A Summary of TOC Developments," Product Technical Note, O.I. Corp., College Station, Texas, 1985.

Berson, S. A. and Yallow, R. S., "Quantititive Aspects of The Reaction Between Insulin and Insulin-Binding Antibody," J. Clin. Invest., **38**, 1996(1959).

Braman, R. S., "Gas Chromatographic Analysis in Air Pollution," in *Chromatographic Analysis of The Environment*, ed. Grob, R. L., Marcel Dekker, New York, 1983.

Bronowski, J., *The Ascent of Man*, Little, Brown and Co., Boston, 1973.

Carr, D. E., *Death of the Sweet Waters*, W.W. Norton, New York, 1966, Chapter 1.

Carson, R., *Silent Spring*, Houghton-Mifflin, Boston, 1962.

Chen, K., Y., "Chemistry of Sulfur Species and Their Removal From Water Supply," in *Chemistry of Water Supply, Treatment, and Distribution*, ed. A. J. Rubin, Ann Arbor Science Publishers, Ann Arbor, 1974.

Clark, L. C., "Polarographic Oxygen Electrode," *American Society of Artificial Internal Organs*, **2**, 41(1956).

Cohn, M. M. and Metzler, D. F., "The Pollution Fighters," N.Y. State Department of Health, Health Education Services, Inc., 1973.

Conlon, R. D., "Refractive Index Monitor Employing Fresnel's Principle," *Review of Scientific Instruments*, **34**, 1418(1963)

Dixon, J. P., *Modern Methods in Organic Microanalysis*, Van Nostrand, Princeton, N.J., 1968.

Drower, M. S., "Water-Supply, Irrigation, and Agriculture," in *A History of*

Technology, ed. Singer, C., Holmyard, E. J., Hall, A. R., Oxford, London, 1954, Vol. I.

Fergusson, J. E., *Inorganic Chemistry and the Earth*, Pergamon, New York, 1982.

Fitchett, A. W., "Analysis of Rain by Ion Chromatography," in *Sampling and Analysis of Rain*, ASTM STP 823, ed. Campbell, S. A., American Society for Testing and Materials., 1983, pp. 29–40.

Freifelder, D., *Physical Biochemistry*, W. H. Freeman, New York, 1982.

Golterman, H. L., Clymo, R. S., Ohnstad, M. A. M., *Methods For Physical & Chemical Analysis of Fresh Waters*, IBP Handbook #8, Blackwell Scientific Pubs., Oxford, 1978.

Graedel, T. E., "Atmospheric Photochemistry," in *The Handbook of Environmental Chemistry*, Vol. 2, Part A, ed. Hutzinger, O., Springer-Verlag, New York, 1980.

Green, D. W. and Reedy, G. T., "Matrix Isolation Studies," in *Fourier Transform Infrared Spectroscopy*, ed. Ferraro, J. R. and Basile, L. J., Volume I, Academic Press, New York, 1978.

Guggenheim, E. A., *Thermodynamics*, North Holland Publishing Co., Amsterdam, 1950.

Guiliany, B. E., "Gas Chromatography in Water Pollution," in *Chromatographic Analysis of The Environment*, ed. Grob, R. L., Marcel Dekker, Inc., New York, 1983.

Hamilton, R. J. and Sewell, P. A., *Introduction to High Performance Liquid Chromatography*, 2ed., Chapman and Hall, New York, 1982.

Hasler, A. D., "Man-Induced Eutrophication of Lakes," in *Global Effects of Environmental Pollution*, ed. Singer, S. F., Springer-Verlag, New York, 1970.

Hatfield, W. D., "Sewage Pollution in the United States is Appalling," *Outdoor America*, May, 1927.

Hauk, R. D., "Atmospheric Nitrogen: Chemistry, Nitrification, Denitrification, and their Interrelationships," in *Handbook of Environmental Chemistry*, Vol. 1, Part C, ed. Hutzinger, O., Springer-Verlag, New York, 1984.

Horne, R. A., *The Chemistry of Our Environment*, Wiley, New York, 1978.

Horne, R. A., *Marine Chemistry*, Wiley, New York, 1969

Huber, S. J. and Hock, B., "A Solid Phase Immunoassay for Quantitative Determination of Terbutryn," *Pfl. Krankh. Pfl.-Schutz*, **92**, 147(1985)

Hudson, L. and Hay, F. C., *Practical Immuunology*, Blackwell Scientific, London, 1980.

Hunt, G. T., "Thin Layer Chromatographic Analysis in Water Pollution," in *Chromatographic Analysis of The Environment*," ed. Grob, R. L., Marcel Dekker, Inc., New York, 1983.

Jolly, W. L., *Modern Inorganic Chemistry*, McGraw-Hill, New York, 1984.

Jolly, W. L., *The Inorganic Chemistry of Nitrogen*, W. A. Benjamin, New York, 1964.

Klotz, I. M., *Chemical Thermodynamics*, Prentice Hall, New York, 1950.

Kolthoff, I. M., Sandell, E. B., Meehan, E. J., and Bruckenstein, S., *Quantitative Chemical Analysis*, Macmillan, London, 1969.

Kortum, G., *Reflectance Spectroscopy*, Springer-Verlag, New York, 1969.

Krieger, J., *Chemical and Engineering News*, March 30, 1987, p. 22.

Lakshminarayanaiah, N., *Membrane Electrodes*, Academic Press, New York, 1976.

Langhorst, M. L., "Photoionization Detector Sensitivity of Organic Compounds," *J. Chrom. Sci.*, **19**, 98(1981)

Lehninger, A. L., *Biochemistry*, Worth, New York, 1970.

Leichnitz, K., *Detector Tube Handbook*, Draeger AG, Lubeck, West Germany, 1985.

Lewis, H.R., *With Every Breath You Take*, Crown Publishers, New York, 1965.

Lingane, J. J. *Electroanalytical Chemistry*, 1st Ed., Interscience, New York, 1953.

Ma, T. S., and Rittner, R. C., *Modern Organic Elemental Analysis*, Marcel Dekker, Inc., New York, 1979.

Meyer, B., "Elemental Sulfur", in *Inorganic Sulfur Chemistry*, ed. Nickless, G., Elsevier, Amsterdam, 1968.

Michaelis, L. and Menten, M. L., "Kinetics of Invertase Action," *Zeitschrift für Biochemie*, **49**, 333 (1913)

Miller, J. M., *Chromatography: Concepts and Contrasts*, Wiley, New York, 1988.

Mindrup, R., "The Analysis of Gases and Light Hydrocarbons by Gas Chromatography," *J. Chrom. Sci.*, **16**, 380(1978)

Moeller, T., *Inorganic Chemistry*, Wiley, New York, 1952

Moelwyn-Hughes, E. A., *Physical Chemistry*, Macmillan, New York, 1961.

Molina, M. J. and Rowland, F. S., "Stratospheric Sink for Chlorofluoromethanes. Chlorine Atom-Catalyzed Destruction of Ozone," *Nature*, **249**, 810(1974)

Moore, W. J., *Physical Chemistry*, Prentice-Hall, New York, 1950

Morris, D. A. N., Smith, M. D., and Greyson, J., "Surface Immune Precipitation, A New Method for Rapid Quantitative Antigen Analysis," *J. Immuno. Methods*, **9**, 363-372 (1976)

Mullins, T., "The Chemistry of Water Pollution," in *Environmental Chemistry*, ed. Bockris, J., Plenum, New York, 1977.

Nyberg, B., "Tough Activists Force Stockmen to Address Animal-Cruelty Issue," Denver Post, December 6, 1987.

O'Brien, M. J., "Liquid Chromatographic Analysis in Air Pollution," in *Chromatographic Analysis of The Environment*, Marcel Dekker, Inc., New York, 1983.

Olsen, E. D., *Modern Optical Methods of Analysis*, McGraw-Hill, New York, 1975.

Pauling, L., *The Nature of the Chemical Bond*, Cornell University Press, Ithaca, N.Y., 1948.

Peters, D. G., Hayes, J. M., and Hieftje, G. M., *A Brief Introduction to Modern Chemical Analysis*, W. B. Saunders, Philadelphia, 1976.

Pickering, W. F., *Modern Analytical Chemistry*, Marcel Dekker, Inc., New York, 1971.

Purnell, J. H., "Correlation of Separating Power and Efficiency of Gas Chromatographic Columns," *J. Chem. Soc.*, 1268(1960)

Rabinowicz, E., *An Introduction to Experimentation*, Addison-Wesley, Menlo Park, CA, 1970.

Robinson, R. A. and Stokes, R. H., *Electrolyte Solutions*, Academic Press, New York, 1959.

Robyt, J. F. and White, B. J., *Biochemical Techniques, Theory and Practice*, Brooks/Cole Pub. Co., Montery, Cal., 1987.

Roth, H., *Quantitative Organic Microanalysis of Fritz Pregl*, (Translated by E. B. Daw), Blackston, Philadelphia, 1937.

Russow, J., "Fluorocarbons," in *The Handbook of Environmental Chemistry*, Vol. 3, Part A, ed. Hutzinger, O., Springer-Verglag, New York, 1980.

Ruzicka, J. and Hansen, E. H., *Flow Injection Analysis*, Wiley, New York, 1981.

Schenk, P. W. and Steudel, R., "Oxides of Sulfur," in *Inorganic Sulfur Chemistry*, ed. G. Nickless, Elsevier, New York, 1968.

Schwalbe-Fehl, M., "Immunoassays in Environmental Analytical Chemistry," *Intern. J. Environ. Anal. Chem.*, **26**, 295(1986).

Seinfeld, J. H., *Air Pollution*, McGraw-Hill, New York, 1975.

Shamel, R. E. and Chow, J. J., *Genetic Engineering News*, July 1987, p. 22.

Shelley, P. E., "Sampling of Water and Wastewater," EPA 600/4-77-039, Environmental Research Information Center, Office of Research and Development, U.S. Environmental Protection Agency, Cincinnati, Ohio, 45268.

Sidgwick, N. V., *The Chemical Elements and Their Compounds*, Oxford, London, 1950

Smil, V., *Carbon-Nitrogen-Sulfur: Human Interference in Grand Biospheric Cycles*, Plenum, New York, 1985.

Smythe, L.E., "Analytical Chemistry of Pollutants," *Environmental Chemistry*, ed. Bockris, J. O'M., Plenum, New York, 1977.

Symanski, J. S. and Bruckenstein, S., "Conductimetric Sensor for Parts per Billion Sulfur Dioxide Determination," *Anal. Chem.* **58**, 1771–1777(1986).

Walton, H. F., "Liquid Chromatographic Analysis in Water Pollution," in *Chromatographic Analysis of The Environment*, ed. Grob, R. L., Marcel Dekker, Inc., New York, 1983.

Wetzel, R. A., Pohl, C. A., and Riviello, J. M., "Ion Chromatography," in *Inorganic Chromatographic Analysis*, ed. J. C. MacDonald, Wiley, New York, 1985.

Yallow, R. S. and Berson, S. A., "General Principles of Radioimmunoassay," *AEC Symp. Series 13, Radioisotopes in Medicine in Vitro Studies*, Oak Ridge, Tenn., 1968.

Youden, W. J., *Statistical Methods for Chemists*, Wiley, New York, 1951

Zelinski, S. G. and Hunt, G. T., "Thin Layer Chromatographic Analysis in Air Pollution," in *Chromatographic Analysis of The Environment*, Marcel Dekker, Inc., New York, 1983.

Zurer, S., "Chemists Solve Key Puzzle of Antarctic Ozone Hole," C&E News, November 30, 1987, p. 25.

Index

Absorbance 147
Acetyl CoA, Coenzyme A 73-76
Acid mine drainage 97
Acid-base indicators 265, 266
Acid-base titrations 255-265
Acid rain 11, 12, 43, 96, 114,
    116, 202, 251, 293
Acidity 84, 194, 273, 274
Acids and bases 255-259
Activated sludge 99, 100
Adenosine triphosphate (ATP) 68
Adsorption chromatography 215, 239
Aeration 99, 100
Air movement 102
Air pollution control district 7, 9
Air-segmented flow 288
Air-water interface 83, 88, 91, 94
Alkalinity 84, 274
Alkylperoxynitrates 117
Alumina 226, 240, 247
American Society for Testing and
    Materials (ASTM)
  ASTM methods
    acidity and alkalinity, measure-
      ment of 273

[American Society for Testing and
    Materials (ASTM)]
  ammonia, determination of 150,
    196
  hydrogen sulfide, determination
    of 165
  Kjeldahl nitrogen, determina-
    tion of 275
  nitrates, determination of 150
  nitrogen dioxide, determination
    of 151
  recommendations for use of gas
    detection tubes 278
  in sampling methods 129
  sulfate, determination of 168,
    254
  sulfides in water, determination
    of 276
  sulfite, determination of 275
  sulfur dioxide, determination of
    151, 275, 276
Directory of testing laboratories
    322
Amino acids 19, 42, 65, 70-73, 79,
    94

329

DATE DUE

NOV 14 2000			
DEC 05 2000			
DEC 05 2000			
DEC 03 2001			
JAN 18 2005			
DEC 17 2007			
			Printed in USA